AF905147

Background and Future Prospects in Life Cycle Assessment

LCA Compendium - The Complete World of Life Cycle Assessment

Series Editors

Walter Klöpffer, LCA Consult & Review, Frankfurt am Main, Germany
Mary Ann Curran, BAMAC, Ltd., LCA & Sustainability Consultant, Cincinnati, OH, USA

Aims and Scope

Life Cycle Assessment (LCA) has become the recognized instrument to assess the ecological burdens and human health impacts connected with the complete life cycle (creation, use, end-of-life) of products, processes and activities, enabling the assessor to model the entire system from which products are derived or in which processes and activities operate. Due to the steady, world-wide growth of the field of LCA, the wealth of information produced in journals, reports, books and electronic media has made it difficult for readers to stay abreast of activity and recent developments in the field. This led to the realization of the need for a comprehensive and authoritative publication.

The *LCA Compendium* Book Series will discuss the main drivers in LCA (SETAC, UNEP/SETAC Life Cycle Initiative, etc.), the strengths and limitations of LCA, the LCA phases as defined by ISO standards, specific applications of LCA, Life Cycle Management (LCM) and Life Cycle Sustainability Assessment (LCSA). Further volumes, which are closely related to these themes will cover examples of exemplary LCA studies ordered according to the importance of the fields of application. They will also present new insights and new developments and will keep the whole work current. The aim of the series is to provide a well-structured treatise of the field of LCA to give orientation and guidance through detailed descriptions on all steps necessary to conduct an LCA study according to the state of the art and in full agreement with the standards.

The *LCA Compendium* Book Series anticipates publishing volumes on the following themes:

- Background and Future Prospects in Life Cycle Assessment
- Goal and Scope Definition in Life Cycle Assessment
- Life Cycle Inventory Analysis (LCI)
- Life Cycle Impact Assessment (LCIA)
- Interpretation, Critical Review and Reporting in Life Cycle Assessment
- Applications of Life Cycle Assessment
- Special Types of Life Cycle Assessment
- Life Cycle Management (LCM)
- Life Cycle Sustainability Assessment (LCSA)
- Life Cycle Assessment Worldwide

For further volumes:
http://www.springer.com/series/11776

Walter Klöpffer
Editor

Background and Future Prospects in Life Cycle Assessment

Springer

Editor
Walter Klöpffer
LCA Consult and Review
Frankfurt am Main
Germany

ISSN 2214-3505
ISSN 2214-3513 (electronic)
ISBN 978-94-017-8696-6
ISBN 978-94-017-8697-3 (eBook)
DOI 10.1007/978-94-017-8697-3
Springer Dordrecht Heidelberg New York London

Library of Congress Control Number: 2014933434

© Springer Science+Business Media Dordrecht 2014
This work is subject to copyright. All rights are reserved by the Publisher, whether the whole or part of the material is concerned, specifically the rights of translation, reprinting, reuse of illustrations, recitation, broadcasting, reproduction on microfilms or in any other physical way, and transmission or information storage and retrieval, electronic adaptation, computer software, or by similar or dissimilar methodology now known or hereafter developed. Exempted from this legal reservation are brief excerpts in connection with reviews or scholarly analysis or material supplied specifically for the purpose of being entered and executed on a computer system, for exclusive use by the purchaser of the work. Duplication of this publication or parts thereof is permitted only under the provisions of the Copyright Law of the Publisher's location, in its current version, and permission for use must always be obtained from Springer. Permissions for use may be obtained through RightsLink at the Copyright Clearance Center. Violations are liable to prosecution under the respective Copyright Law.
The use of general descriptive names, registered names, trademarks, service marks, etc. in this publication does not imply , even in the absence of a specific statement, that such names are exempt from the relevant protective laws and regulations and therefore free for general use.
While the advice and information in this book are believed to be true and accurate at the date of publication, neither the authors nor the editors nor the publisher can accept any legal responsibility for any errors or omissions that may be made. The publisher makes no warranty, express or implied, with respect to the material contained herein.

Printed on acid-free paper

Springer is part of Springer Science+Business Media (www.springer.com)

Foreword

The regulatory approaches to control point-source releases of waste that were used in the 1970s and 1980s were instrumental in creating a change for improved environmental management. However, because these were based on a single stage in the product's life cycle (e.g., production, use, or disposal), or a single issue (e.g., individual chemical limits), they did not always lead to a net environmental benefit.

With the improvement in the treatment of air and water emissions and waste from manufacturing operations, there was recognition that end-of-pipe treatment can only go so far. Additional examination of what enters the end-of-pipe treatment led to the development of pollution prevention and pre-treatment programs. These programs were influenced by a combination of government regulations and organizations who realized that they could save money by preventing pollution. 3M's Pollution Prevention Program (3Ps), for example, was a landmark program initiated in the 1980s which has saved 3M nearly one billion USD since the 1980s. As well, it has prevented emissions and waste from entering the environment.

At the same time, society realized that attempting to solve our environmental problems went beyond the manufacturing facilities. We realized that products (and packaging) were creating an enormous amount of solid waste after their useful life was over. We also learned that how much electricity used in the use stage of a product influenced the amount of mercury being released into the environment by coal power plants, to provide another example. Many other observations were occurring by organizations and leaders who were asking whether there were other ways to more completely understand that full impact of society on the planet. Life cycle assessment (LCA) surfaced as one of those emerging tools that was advanced enough to fill the gap to allow us to more fully understand the risks, opportunities, and trade-offs among the many stages of product system life cycle and the multiple impacts that could occur at each stage.

In 1990, the Society of Environmental Toxicology and Chemistry (SETAC) sponsored an international workshop (in Smugglers Notch, Vermont, USA) where the term 'life cycle assessment' was coined. SETAC, the International Organization for Standardization (ISO), and the United Nations Environment Programme (UNEP) have since established and advanced the understanding and use of the LCA framework, methodology, and data.

LCA, as governed by the ISO standards 14040 and 14044, has become a recognized instrument to assess the ecological burdens and human health impacts connected with the complete life cycle (creation, use, end-of-life) of products, processes and activities, enabling the practitioner to model the entire system from which products are derived or in which processes and activities operate.

We have seen a dramatic shift in the development and application of LCA over that last 20 years, as part of the modern era of LCA. Our initial efforts were to develop and enhance basic methodological elements, e.g., goal and scope definition, inventory analysis, impact assessment, and data quality and availability. These efforts are continuing. However, over the last 5–8 years, we have seen an increase in the demand for life cycle information as shown by green building, retail, electronic, and purchasing expectations. With this demand there is the need to make the life cycle information available to non-LCA specialists in professional functions, like procurement, innovation, marketing, etc. While LCA will continue to be a tool for specialists, life cycle information will be embedded in existing decision support tools (e.g., CAD systems) and business practices (e.g., the stage gate process). Mainstreaming of life cycle information is upon us and its use will grow steadily to allow us to speed and scale up the transition to more sustainable product systems.

This book series on LCA is the first work of its kind—a major undertaking to create a comprehensive collection of writing aimed at illuminating all aspects of LCA. Volumes in the series will discuss such topics as the main drivers in LCA (SETAC, UNEP/SETAC Life Cycle Initiative, etc.), the strengths and limitations of LCA, the gaps and research needs in LCA, LCA phases as defined by ISO standards, specific applications of LCA, Life Cycle Management (LCM) and Life Cycle Sustainability Assessment (LCSA). Written by international LCA experts, this book series will be an invaluable resource for those involved in assessing environmental performance through all stages of goods and services.

I encourage you to read, learn and apply the information and knowledge in the chapters and volumes as they are released. Use them as a reference to apply life cycle assessment frameworks, methodologies, information, data and insights. Let's work together to mainstream the use of life cycle information to improve the sustainability of product systems.

James A. Fava
PE INTERNATIONAL, Inc., West Chester, PA19380, USA

Acknowledgments

The *LCA Compendium* Book Series complements *The International Journal of Life Cycle Assessment*, which has been published by Springer since 2008. Since this is the first volume of the *LCA Compendium*, we first acknowledge and appreciate that Springer accepted to publish this complementary work, the first of its kind in the field of Life Cycle Assessment. In particular we thank Paul Roos (Environmental Sciences) for recognizing the possibility of such an extensive publication and for his guidance in bringing it to life. We also thank Betty van Herk for organizing the preparation of the manuscripts.

Next, we would like to thank all the volume editors who have committed to editing further volumes and the many authors who agreed to write specifically defined chapters on all important issues related to LCA.

Our next thank you is to Almut B. Heinrich, former managing editor of *The International Journal of Life Cycle Assessment*, who is acting as the managing editor for this book series.

Turning to this volume, we thank our fellow contributors: James Fava, Matthias Finkbeiner, Guido Sonnemann, Sonia Valdivia, Almut B. Heinrich, and the many co-authors, first for their clear-sighted willingness to cooperate on this new book series and second, for devoting their time in sharing their expertise and experience in the individual chapters.

Walter Klöpffer and Mary Ann Curran
Series Editors—*LCA Compendium*

Contents

Contributors

Robert Ackermann Department of Environmental Technology, Chair of Sustainable Engineering, Technische Universität Berlin, Berlin, Germany

Vanessa Bach Department of Environmental Technology, Chair of Sustainable Engineering, Technische Universität Berlin, Berlin, Germany

Markus Berger Department of Environmental Technology, Chair of Sustainable Engineering, Technische Universität Berlin, Berlin, Germany

Gerhard Brankatschk Department of Environmental Technology, Chair of Sustainable Engineering, Technische Universität Berlin, Berlin, Germany

Ya-Ju Chang Department of Environmental Technology, Chair of Sustainable Engineering, Technische Universität Berlin, Berlin, Germany

Mary Ann Curran BAMAC Ltd., Cincinnati, Ohio, USA

James A. Fava PE INTERNATIONAL, Inc., West Chester, PA, USA

Matthias Finkbeiner Department of Environmental Technology, Chair of Sustainable Engineering, Technische Universität Berlin, Berlin, Germany

Marina Grinberg Department of Environmental Technology, Chair of Sustainable Engineering, Technische Universität Berlin, Berlin, Germany

Almut B. Heinrich Heidelberg, Germany

Walter Klöpffer LCA Consult & Review, Frankfurt am Main, Germany

Annekatrin Lehmann Department of Environmental Technology, Chair of Sustainable Engineering, Technische Universität Berlin, Berlin, Germany

Julia Martínez-Blanco Department of Environmental Technology, Chair of Sustainable Engineering, Technische Universität Berlin, Berlin, Germany

Nikolay Minkov Department of Environmental Technology, Chair of Sustainable Engineering, Technische Universität Berlin, Berlin, Germany

Laura Morrison PE INTERNATIONAL, Inc., Boston, MA, USA

Sabrina Neugebauer Department of Environmental Technology, Chair of Sustainable Engineering, Technische Universität Berlin, Berlin, Germany

René Scheumann Department of Environmental Technology, Chair of Sustainable Engineering, Technische Universität Berlin, Berlin, Germany

Laura Schneider Department of Environmental Technology, Chair of Sustainable Engineering, Technische Universität Berlin, Berlin, Germany

Andrea Smerek PE INTERNATIONAL, Inc., Ottawa, ON, Canada

Guido Sonnemann Project Management Office of the UNEP/SETAC Life Cycle Initiative, Université Bordeaux 1, Talence Cedex, France

Sonia Valdivia Secretariat of the UNEP/SETAC Life Cycle Initiative, United Nations Environment Programme, Paris Cedex 09, France

Kirana Wolf Department of Environmental Technology, Chair of Sustainable Engineering, Technische Universität Berlin, Berlin, Germany

Chapter 1
Introducing Life Cycle Assessment and its Presentation in 'LCA Compendium'

Walter Klöpffer

Abstract This chapter spans the time from the early days of Life Cycle Assessment—LCA (the time of the so-called 'proto-LCAs' between about 1970 and 1990), until recent trends of simplified/streamlined LCAs, the footprint specifications (carbon footprint, water footprint) and Life Cycle Sustainability Assessment—LCSA.

Important benchmarks along this span are the harmonisation of LCA by SETAC (Society of Environmental Toxicology and Chemistry) and the standardisation of LCA by ISO (International Standardisation Organisation).

The basic discussions within SETAC occurred between 1990 and 1993.

The first attempt to develop a suitable LCA-structure was achieved during the SETAC workshop 'A Technical Framework for Life Cycle Assessments' in August 1990, held in Smugglers Notch, Vermont, USA. The LCA-structure, the famous 'SETAC triangle', consisted of three components: Inventory—Impact Analysis—Improvement Analysis.

SETAC revised the framework during the Sesimbra workshop in 1993. It was the merit of SETAC to initiate a standardisation process which culminated in the 'Guidelines for Life-Cycle Assessment: A Code of Practice'. The LCA-structure, again a triangle, now included four components: Goal Definition and Scoping—Inventory Analysis—Impact Assessment—Improvement Assessment.

This structure was only slightly modified by the ISO standardisation process: The fourth phase 'Improvement Assessment' (formerly 'Improvement Analysis') was replaced by 'Interpretation'.

After the harmonisation of LCA by SETAC, the International Standardisation Process was soon initiated (Autumn 1993 in Paris), but it took seven years for the first series of LCA standards to be published (ISO 14040, ISO 14041, ISO 14042, ISO 14043).

The successful first series of ISO LCA standards superseded the SETAC 'Code of Practice', the Nordic guidelines and several national standards and became the uncontested model of an environmental life cycle standard. The series 14040 ff was revised once and condensed into two standards 14040 and 14044 (2006).

The four-phase structure was not altered

W. Klöpffer (✉)
LCA Consult & Review, Am Dachsberg 56E,
60435, Frankfurt am Main, Germany
e-mail: walter.kloepffer@t-online.de

W. Klöpffer (ed.), *Background and Future Prospects in Life Cycle Assessment*,
LCA Compendium – The Complete World of Life Cycle Assessment,
DOI 10.1007/978-94-017-8697-3_1, © Springer Science+Business Media Dordrecht 2014

This chapter discusses the four phases of the LCA-structure by SETAC and ISO which are the subject of four volumes—Goal and Scope Definition in LCA; Life Cycle Inventory Analysis; Life Cycle Impact Assessment; Interpretation, Critical Review and Reporting. The remaining volumes follow a structure outside the ISO-framework: Applications of LCA, Special Types of LCA, Life Cycle Management, and Life Cycle Sustainability Assessment.

Keywords Carbon footprint · Goal and scope definition in LCA · Harmonisation of LCA by SETAC · Improvement analysis · Interpretation · ISO (International Standardisation Organisation) · Life cycle assessment—LCA · Life cycle impact assessment—LCIA · Life cycle inventory analysis—LCI · Life cycle sustainability assessment—LCSA · Proto-LCAs · SETAC (Society of Environmental Toxicology and Chemistry) · SETAC triangle · Simplified LCA · Standardisation of LCA by ISO · Streamlined LCA · Water footprint

1 What is Life Cycle Assessment?

Life Cycle Assessment (LCA) is a science-based, comparative analysis and assessment of the environmental impacts of product systems. It is distinguished from other environmental assessment methods by two constitutive and unique features: the analysis from 'cradle-to-grave' and the 'functional unit'. Together, the application of these features allows the comparison of product systems fulfilling the same, or very similar, purpose. In order to explain the basic principles of LCA and the most important terms, we paraphrase the relevant international guidelines and standards (SETAC 1993; ISO 1997, 2006a). 'Cradle-to-grave' means that all the important steps in the life cycle of a product are included in the analysis, namely the extraction of raw materials from the environment (soil, water, air), the production of materials and the final products, their use and waste removal or recycling. Any transportation that occurs across these steps is also accounted for. The 'products' are defined as 'goods and services' in all relevant standards. The final products in services are intangible but need the same processes, energy sources, etc. as tangible products or goods; the definition of the life cycle has to be modified accordingly case-by-case.

The concept of 'life cycle' used in LCA is always the physical life cycle, rather than the 'marketing life cycle' which starts with planning, R&D and design, introducing a product into the market, producing, selling, leasing, etc. until the product is taken out of the market. This definition can also be used for goods and services where a functional unit can be defined. In an environmental LCA it is hard to confuse the two related terms. If, however, a Life Cycle Costing—LCC[1] is to be added

[1] A pre-guideline or framework of LCSA was published by UNEP/SETAC at the end of 2011. The final version has been published in 2012 at: http://lcinitiative.unep.fr. It contains the three-pillar equation:

LCSA = LCA + LCC + SLCA
(SLCA = Social LCA).

to the LCA, great care has to be taken in order to avoid confusion. The type of LCC compatible with LCA should better be called 'Environmental LCC' to show that it adheres to the physical life cycle (Hunkeler et al. 2008, Klöpffer 2008, Swarr et al. 2011).

The 'functional unit' is the basis of comparison of product systems (goods *and* services) if they provide the same or a very similar function (hence the name). It describes quantitatively the function of the product systems to be compared, e.g. the packaging of 1000 L of a beverage and its transport to the point of sale, recycling and waste removal (Klöpffer and Grahl 2009, 2014). It is usual praxis in LCA to neglect small differences of products (e.g. aesthetic ones), as long as they have no or only minor influence on the environmental impacts of the product system. A clear definition of the product systems to be investigated, including their boundaries with the environment and the rest of the technosphere, is of paramount importance; it has to be provided in the Goal and Scope Definition.

Underpinning all terms introduced so far means understanding that products in the sense of LCA are product *systems* rather than the material product we may use or the service we may hire. Behind these obvious aspects of products to be observed in daily life there is a multitude of upstream and downstream processes[2], intermediate products, transport processes, packaging and energy use, to name just a few. Downstream from the use phase[3] starts the end-of-life (EOL) phase (waste management and recycling). In general, even experienced life cycle practitioners, if confronted with a new problem, are not aware of the full complexity of the 'product tree'[4], the 'supply chain'[5] and the EOL. This has two consequences: first, the construction of the product tree has to be done on the basis of best available information and may require some research; and, second, the system has to be tailored, and small amounts of residual inputs and outputs have to be cut off. Metaphorically speaking, the product tree has to be cut out of the dense network of the technosphere. In comparing two or more products, their system boundaries[6] have to be defined in a similar way.

To bring together all product-related terms, we may call LCA (to be precise: the Life Cycle Inventory—LCI) a simplified system analysis. To visualize such systems, the smallest units for which data are available—the unit processes—are shown as boxes which are connected to other unit processes from which they obtain inputs and to which they transfer substances, materials and energy. Releases into the environment (emissions) leave the system. There are also imports from

---

[2] Upstream processes: toward the 'cradle', downstream processes: toward the 'grave'.

[3] Use phase: e.g. driving a car for a certain time; the use phase is the centre of most life cycles defined in LCA.

[4] Product tree: the most common form of graphical presentation of product life cycles.

[5] Supply chain: usual, but misleading (since suggesting linearity) designation of the upper part of a product tree or branches thereof; modern economies are characterised by a high degree of work-sharing.

[6] The system boundary separates the system to be studied from the rest of the technosphere and the environment.

the environment in the form of oil and coal, gases, minerals, water, radiation from the sun, etc. The systems studied in LCI are part of the technosphere, whereas the environment is receiving the releases and provides inputs at the interface between the two spheres. Specific interactions between environment and technosphere are quantified in an LCA-phase called Life Cycle Impact Assessment—LCIA and discussed in the final phase Interpretation.

The success of LCA as an environmental assessment tool lies in its broad applicability, to all product systems for which data can be generated and the possibility to compare the results with competing (or improved) systems. This offers a great opportunity for improving products and the technosphere (thus improving the environment), but it has also led to producing false or at least exaggerated environmental product claims. The misuse potential was especially high in the time of the 'proto-LCAs' (see Sect. 2) and led to the harmonisation of the different early methods by SETAC and finally to the standardisation by ISO (Klöpffer 2006, 2012).

2 LCA—How it Came About

2.1 The Early Time

The first LCAs—the author called them 'proto-LCAs' (Klöpffer 2006)—were performed between about 1970 and 1990, when the harmonisation of the early life cycle methods developed in North America and Europe started (Fava et al. 1991). Even these proto-LCAs presented the two most important features discussed in Sect. 1: system comparison by functional unit and cradle-to-grave analysis. They consisted of a life cycle inventory (LCI) and sometimes a rudimentary form of impact analysis. The emphasis of the early LCAs was on energy saving and resource conservation rather than on pollution. This can be seen in the designation of the proto-LCAs chosen by Franklin Associates Ltd (FAL): Resource and Energy Profile Analysis (REPA) (Hunt et al. 1992). The history of this method dates back to the late 1960s when Franklin and Hunt worked at Midwestern Research Institute (MRI), as related by the two pioneers in an invited personal account in the first issue of Int J Life Cycle Assess (Hunt and Franklin 1996). According to this story, the first REPA study was performed for the Coca Cola Company; the commissioner of the study, Harry Teasley, evidently contributed to the method development with ideas of his own. FAL has performed hundreds of REPAs in the following decades and still exists as division of a larger consultancy (ERG).

The first Franklin-type proto-LCA in Europe was performed in the early 1970s at Battelle Institut e.V. (Frankfurt am Main) about the comparative assessment of beverage containers (Oberbacher et al. 1996; Oberbacher 1975). The inventory ('proto-LCI') was already well developed, and even an economic analysis, a kind of life cycle costing (LCC) was added to this life cycle study which was classified by the authors as a system analysis without a name of its own. The importance of this work was recognised at once but did not immediately lead to additional research contracts … until 1988 when it helped the author, who was only marginally involved in the original study, to get his first LCA project (Klöpffer 1989). It finally opened the door to an international cooperation with Battelle Memorial Institute (Columbus, Ohio, USA) and the Society of Environmental Toxicology and Chemistry (SETAC).

It should be noted that the first LCA PhD dissertation in Germany was performed at the Technical University Berlin (Franke 1983), later to become one of the centres of LCA research. Related activities occurred and are still going on at the University of Stuttgart which can be considered the cradle of PE International.

In England, Ian Boustead started related life cycle work in the early 1970's (Boustead 1996), again using bottles for liquids (milk) as first study objects. His work rapidly extended into industrial systems, where the scarcity of data soon turned out as a bottle neck. He started a huge data collection effort which culminated in a first book (Boustead and Hancock 1979) on data useful for extending Life Cycle Thinking into a quantitative analysis. Another lasting result of this pioneering work has been the creation of a vast commercial data bank for LCA practitioners. With the advent of modern LCA (Sects. 2.2 and 2.3), numerous LCI data collections for commodities, especially for polymers, were created by Boustead (e.g. Boustead 1993). Similar to other pioneers who approached LCA from an engineering perspective, he never supported or practiced life cycle impact assessment methodology.

The rôle of Switzerland as a leading LCA power was also founded in the 1970's (Fink 1997). The leading organisations at that time were EMPA[7] and the University of St. Gallen. The proto-LCAs were called 'Ökobilanz' and this is still the official German name for LCA. As in the UK, data collection for the most important packaging materials was initiated by the Swiss Federal Office for the Environment (BUS 1984). The update (Habersatter and Widmer 1991) was of great influence in the whole German-speaking LCA community. It already contained a rudimentary impact assessment.

Ruedi Müller-Wenk tried to improve the industrial praxis (economy) by inclusion of environmental protection and life cycle thinking (Müller-Wenk 1978). The combination of the different strings of thought and practice ultimately led to the typical Swiss ecobalance method in which the different weighted impacts of product systems are expressed in one figure (Ahbe et al. 1990). This method is still used, after some modifications, in Switzerland (Frischknecht et al. 2009).

The early time of LCA in France is nicely presented by (Blouet and Rivoire 1995). The leading consultant, Écobilan in Paris, contributed early and significantly to the big problem of most inventories: allocation (Heintz and Baisnée 1992; Huppes and Schneider 1994; Ekvall and Tillman 1997; Heijungs and Frischknecht 1998; Curran 2008).

Japan entered the LCA arena around 1990 (Finkbeiner and Matsuno 2000a, b). A biannual series of 'Ecobalance conferences' made the LCA research and praxis done in this country known in the world. As in other countries, a national data bank

[7] Eidgenössische Materialprüfungs- und -forschungsanstalt.

was created as a fundament for the steep increase of the use of LCA around the year 2000.

Last, but not least, two pioneers from Scandinavia have to be mentioned: Gustav Sundström and Allan Astrup Jensen. The former, a civil engineer from Malmö (SE), pioneered the early LCAs of carton packaging (Tetra Pak) (Lundholm and Sundström 1985). Jensen (DK) was strongly involved in the transition from the proto-LCA phase to the harmonised and later standardised LCA. Being convinced that the application of LCAs in industry is more important than academic research about it, he later became the founder of Life Cycle Management—LCM (see Sect. 4.3) as a discipline of its own (Jensen 2007).

James A. (Jim) Fava of the United States has played a leadership role in LCA development at the global level. He was a key promoter of the work conducted under the auspices of SETAC and the early technical workshops on LCA, primarily in the years 1990 through 1993. Later, he was instrumental in forging the UNEP and SETAC alliance to build the UNEP/SETAC Life Cycle Initiative (2002).

2.2 Harmonisation by SETAC

The role of SETAC in the development of LCA is described by Jim Fava in this volume, Chapter 2, and was dealt with earlier by the author to honour Helias Udo de Haes on the occasion of his retirement from CML Leiden (Klöpffer 2006). Although the SETAC structure was slightly modified during the standardisation by ISO, the basic discussions occurred in this scientific society in a relatively short period between 1990 and 1993. The author called this time the 'heroic time of SETAC' (with regard to LCA; see also Jensen and Postlethwaite 2008). SETAC has continued to offer a discussion forum for LCA and related life cycle methods. Together with the United Nations Environment Programme (UNEP), a joint project, the 'UNEP/SETAC Life Cycle Initiative' was founded 2002 (Töpfer 2002) leading to a globalisation of the life cycle methods, above all LCA, Life Cycle Management—LCM (Remmen et al. 2007) and most recently Life Cycle Sustainability Assessment—LCSA (Valdivia et al. 2011). The UNEP/SETAC Life Cycle Initiative is presented in this volume, Chapter 4.

Whereas the proto-LCAs essentially consisted of a life cycle inventory and—frequently, but not always—a rudimentary impact assessment, modern LCAs consist of four components. The first attempt to find a suitable structure was done during the SETAC workshop 'A Technical Framework for Life Cycle Assessments' in August 1990, Smugglers Notch, Vermont (Fava et al. 1991).

This structure, the famous 'SETAC triangle' (Fig. 1.1) consisted of three components:

- Inventory
- Impact Analysis
- Improvement Analysis

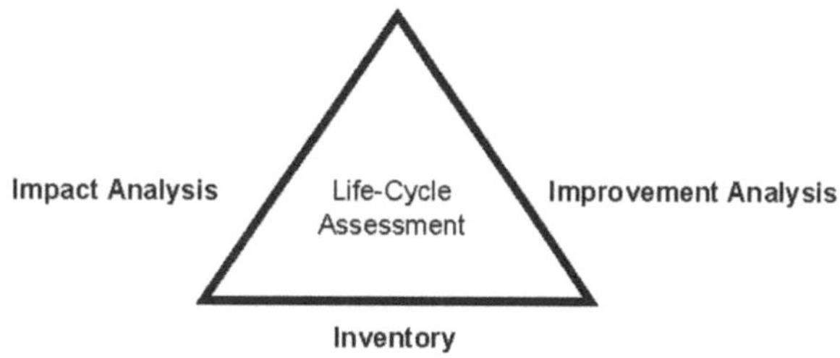

Fig. 1.1 The famous 'SETAC triangle' 1991 (Fava et al. 1991, p. 1)

As stated in the text, most LCAs at that time (typical 'proto-LCAs', see Sects. 1 and 2.1) consisted of inventories only. The component 'Improvement Analysis' was invented during the workshop and experienced a stormy fate in the years to come.

The naming of the 'new' method resulted in Life-Cycle Assessment (LCA). This designation superseded different names used for the proto-LCAs such as life cycle analysis (Ciambrone 1997), ecobalance, REPA (Sect. 2.1) and was taken over by the International Standard Organisation—ISO[8] and The International Journal of Life Cycle Assessment (Sect. 3). The book by Ciambrone, an industry manager and consultant, nicely describes the proto-LCA, many basic techniques of which are still valid today, but neglects any new developments brought about by the harmonisation pioneered by SETAC (Klöpffer 1998a). The book appeared at about the same time when the first modern books on LCA were printed (Curran 1996; Wenzel et al. 1997; Hauschild and Wenzel 1998).

SETAC finally came up with a revised structure during the Sesimbra workshop, immediately after the first SETAC World Conference in Lisbon March/April 1993. The little fishing and recreation village south of Lisbon was to become the birthplace of standardised Life Cycle Assessment: already in August 1993, the 'Guidelines for Life-Cycle Assessment: A Code of Practice' (SETAC 1993) appeared based on the input by 47 invited experts.

The LCA structure, again a triangle (Fig. 1.2), now included four components:

1. Goal Definition and Scoping
2. Inventory Analysis (Materials and Energy Acquisition; Manufacturing; Use; Waste management)
3. Impact Assessment (Ecological Health; Human Health; Resource depletion)
4. Improvement Assessment

How important the inclusion of the (new) first component was, turned out during the ISO process (Sect. 2.3) and after considering the scientific standing of LCA (Sect. 7). Since the second component 'Inventory Analysis' was already well developed in the pre-SETAC time, the new structure focused on 'Impact Assessment' that defined three steps:

- Classification (assigning the data from the inventory table to impact categories)
- Characterisation (aggregation of impacts within the impact categories)

[8] Without hyphen (ISO 1997, 2006a).

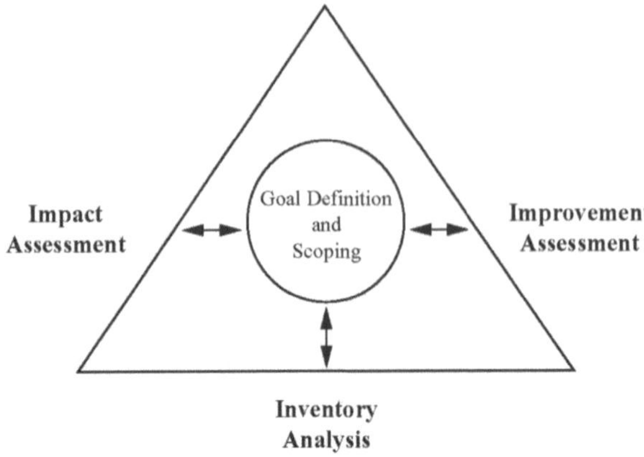

Fig. 1.2 LCA structure in the SETAC 'Code of Practice' of 1993. p 11

- Valuation (weighting of impact results in case of unclear situations, e.g. no clear 'winner' in a comparative LCA)

The terminology used had its origin in a very influential Dutch report which appeared in English in Spring 1993 (Heijungs et al. 1993). Further workshops between Smugglers Notch (1990) and Sesimbra (1993) formed the structure and, of course, the content of LCA.

With commendable foresight, the workshop attendants and editors included the description of a 'peer review' process which should be performed for each LCA:

1. "The peer review process enhances the scientific and technical quality of LCAs; and
2. the process helps to focus study goals, data collection, and provides a critical screening of study conclusions, thereby enhancing study credibility".

Furthermore, an interactive and accompanying peer review was preferred to an 'a posteriori' review for good reasons (Klöpffer 2012).

The success of the 'code of practice' was immediate, and it is not exaggerated to say that it served as 'blue print' for the standardisation by ISO.

2.3 Standardisation by ISO

After harmonisation of LCA by SETAC, the international standardisation process was soon initiated (Autumn 1993 in Paris)[9], but it took seven years until the first series of LCA standards was finished (ISO 1997, 1998, 2000a, 2000b). During that time, the Scandinavian countries prepared LCA guidelines by the 'Nordic Council'

[9] Personal communication by Dr. Manfred Marsmann, chair of ISO/TC 207 'Environmental Management', SC 5 'Life Cycle Assessment', see also (Marsmann 2000).

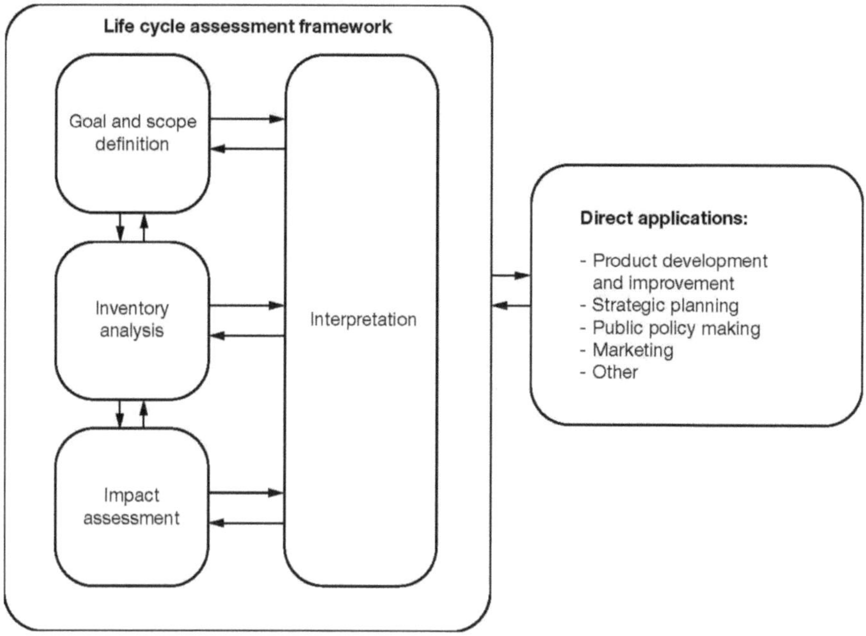

Fig. 1.3 Phases of an LCA (ISO 14040, Fig. 1)

(Lindfors et al. 1995). This and continuing LCA activities by SETAC provided input into the standardisation process.

Fortunately for the LCA community, the first international standard (ISO 14040) dealing with the principles and framework appeared in print already in 1997 (Marsmann 1997). It is this basic LCA standard, later slightly updated (ISO 2006a), which became the 'mother' of all life-cycle based standards and defined the now generally accepted structure of LCA (Fig. 1.3).

The only major change compared to the last scheme by SETAC (see Fig. 1.2) consists in the replacement of 'Improvement Assessment' by 'Interpretation' (see Fig. 1.3). One reason of this retreat behind the ambitious SETAC aim was certainly the fear by industry that an improvement assessment may become obligatory for all LCAs. Another reason for the change is the fact that an LCA may indeed serve different purposes, not only product improvement. Some uses of LCA are given outside the LCA frame in a non-exhaustive list of 'Direct applications':

- Product development and improvement
- Strategic planning
- Public policy making
- Marketing
- Other

The author would add 'teaching and learning' (about environmental burdens connected with product systems) high up in the list as a most useful application of LCA.

A further much commented feature of the ISO framework (already indicated in Fig. 1.2, but only with regard to G&SD) consists in the double arrows symbolizing that the phases may be modified if new aspects emerge during the performance of an LCA. Clearly, a new impact category in LCIA may require a more detailed LCI. Even the goal and scope definition (G&SD) can be altered, but only in written form. This gives the necessary flexibility in performing LCAs and hinders, at the same time, arbitrary changes in G&SD.

Two other aspects have to be considered with regard to the structure of LCA:

- An LCA has to consist of *all* stages. An LC study without Life Cycle Impact Assessment (remember the proto-LCAs!) is a Life cycle inventory study and consists of G&SD, LCI and Interpretation.
- An LCIA has to consist of a *set* of impact categories, one is *not* enough. This means that so called 'Carbon footprint' and other 'footprint' studies with only one impact category are *not* LCAs. Actually they contradict the spirit of LCA fundamentally.
- The norm 14040 has to (*"shall"*) be used in conjunction with ISO 14044, not stand alone and not with another set of requirements (Finkbeiner et al. 2006).
- A 'critical review' (this term superseded the 'peer review' proposed by SETAC) is obligatory for LCA studies to be used for 'comparative assertions' and is recommended, but voluntary, for all other LCAs. The prevention of misuse of the method is high up in the agenda of ISO 14040 ff (Klöpffer 2012).

The very successful first series of ISO LCA standards (ISO 1997, 1998, 2000a, b) superseded the SETAC 'Code of practice' (SETAC 1993), the Nordic guidelines and several national standards as well and became the uncontested model of an environmental life cycle standard. The series 14040 ff was revised once and condensed into two standards 14040 (ISO 2006a) and 14044 (ISO 2006b). The structure, which is our main interest in this chapter, was not altered. Furthermore, the two new standards dealing with the principles and framework of LCA (14040:2006) and with the requirements and guidelines (14044:2006) are coupled together with one "shall" so that the more open standard 14040 cannot be used together with a less demanding set of requirements and still claims to be performed according to ISO. This is extremely important—together with the critical review—for the prevention of misuse of the standards.

2.4 Recent Trends

Whereas the structure of LCA as a science-based method for environmental assessments (presented in Sect. 3) is unchallenged, new developments occurred in recent years (or older ones were rediscovered, as in the case of sustainability assessment) requiring more flexibility and/or detailed requirements than ISO 14040+44 can offer. The changes suggested—and partly are enacted in separate ISO norms (this volume, Chap. 3[10]) or other guidelines—can be classified as belonging to one of the following trends:

- Make it simpler and more flexible
- Reduce the life cycle impact assessment to one impact category
- Expand the environmental LCA to a life-cycle based sustainability assessment

The *first* tendency is as old as modern LCAs; e.g. a SETAC Europe working on simplified LCA came out with definitions and proposals of how to perform such studies—either as a first step to a full LCA or as a stand-alone study (Christiansen 1997). Other studies and workshops also dealt with the problem of 'streamlining' LCA (Curran and Young 1996; Weitz et al. 1996; Hunt et al. 1998; Hochschorner and Finnveden 2003). It seems that the advent of Life Cycle Management—LCM offering a broader palette of methods—the so-called 'tool box' (Wrisberg and Udo de Haes 2002)—took some pressure out of the simplified LCA debate. Furthermore, recent improvements of data bases and LCA software made simplified LCAs feasible at short notice (unfortunately often at the price of a thorough understanding of the systems investigated).

Clearly, any simplified/streamlined LCA studies should be used internally only and upgraded to a full, critically reviewed LCA if a broader public is to be informed (e.g. in marketing) or a scientific publication is planned.

The *second* tendency has become *en vogue* recently, mostly under the name 'footprint', a reminder of a simplified impact assessment long before SETAC defined LCIA. The basic idea was to estimate the area of forest needed to provide the same service as a certain amount of fossil energy. The advantage of such simple parameters is being pictorial. This advantage is lost in modern footprints, the most used being the 'carbon footprint' (CF) which is just another name for the Global Warming Potential (GWP) caused by the Green House Gas (GHG) emissions, mainly carbon dioxide (CO_2), methane (CH_4) and dinitrogen oxide (N_2O). As can be seen, the third gas of this infamous troika does not contain the element carbon (C), and the same is true for sulphur hexafluoride (SF_6), one of the strongest GHG. Despite all deficiencies of the CF concept, we have to deal with it seriously (Finkbeiner 2009) in order to prevent inaccurate calculations of the GWP. An international standard ISO DIS 14067 for CF has been elaborated (ISO 2013), and a British pre-standard PAS 2050 has been available since 2008 (BSI 2008; Sinden 2009). These standards refer to ISO 14040 as basis document and describe in detail how the GWP of products is calculated and how it should be communicated. This goes beyond ISO 14040+44, since the LCA standards do not prescribe for any impact category how the indicators chosen should be determined. So, in the end, CF may enrich LCA, although a stand-alone CF study is not and never will be an environmental LCA.

The same may be true for another 'footprint' which is more innovative and concerns the use of water, especially in countries where water is a scarce resource (Pfister et al. 2009; Berger and Finkbeiner 2010). Water use has been on the list

[10] The international standards as the constitution of LCA: the ISO 14040 series and its offspring by Matthias Finkbeiner.

of life cycle impact categories since the time of LCA harmonisation (Sect. 2.2), but concrete indicators and characterisation factors have not been worked out until recently. This may be due to the fact that LCA has been developed in countries (e.g. Switzerland) in which water scarcity is—in general—no problem at all. The global approach of the UNEP/SETAC Life Cycle Initiative brought water on the LCA agenda. The term 'water footprint' is much used and can be integrated into LCIA—as an impact category belonging to resource use. Water is a renewable abiotic resource and the basis of life on earth; it therefore has many other aspects than as resource for human use.

An 'Ecological footprint' has recently been suggested (SETAC Europe 2011) and may be seen as counterweight against the preponderance of energy—and resource-related impacts in classical LCIAs. This predominance was much more overwhelming, however, in the time of the proto-LCAs (see, e.g., REPA), and the input given by CML[11] in the early 1990s was seen as breakthrough of the environmental sciences in LCA (Gabathuler 1997).

The **third** trend is a come-back of sustainability as the ultimate assessment goal (World Commission on Environment and Development 1987). In the 'three pillar' interpretation of sustainability, environmental, economic and social aspects have to be considered and weighted against each other. In life cycle product assessment, LCA deals with the environmental aspects only (ISO 1997, 2006a). In order to give the full picture, however, an economic and a social life cycle assessment have to be considered with the environmental one (Klöpffer 2003, 2008; Finkbeiner et al. 2010; Valdivia et al. 2012). This concept has been represented as the following non-numerical 'equation':

$$\text{LCSA} = \text{LCA} + \text{LCC} + \text{SLCA}$$

LCSA: Life Cycle Sustainability Assessment
LCA: (environmental) Life Cycle Assessment
LCC: (environmental) Life Cycle Costing
SLCA: Social Life Cycle Assessment

In order to apply this 'equation' properly, it is essential that the system boundaries of the three life cycle assessments are compatible, ideally equal.

With regard to harmonisation and standardisation, (environmental) LCA is most advanced (ISO 2006a, b) and LCSA has been recognised by UNEP/SETAC only recently (Valdivia et al. 2011), albeit after a long discussion at working group level.

The (environmental) LCC[12] has been defined in a 'SETAC Code of practice' (Swarr et al. 2011), based on a more comprehensive treatise relating to the results of a SETAC Europe LCC working group (Hunkeler et al. 2008).

[11] Centrum voor Milieukunde Leiden.

[12] Stand-alone LCC as a purely economic method is older than LCA and has been used to calculate the true costs of long lived products—including the costs of the use and end-of-life phases in addition to the purchase.

The various approaches to Social Life Cycle Assessment (SLCA) are described in a global guideline (UNEP/SETAC 2009; Benoît et al. 2010). It remains to be seen which approach fits best to LCA and LCC. A quantitative approach based on LCI data and statistics (a kind of poorness index) has been proposed by one of the pioneers of sustainability assessment (Hunkeler 2006). Poorness is at the heart of most social problems and seems to play a similar role in SLCA as energy in LCA: as the common root of several impacts.

3 The Structure of LCA According to ISO 14040 and 14044

The structure of LCA, as defined by the international standards, follows the scheme in Fig. 1.3 (Sect. 2.3). It consists of the four phases:

1. Goal and scope definition
2. Inventory analysis
3. Impact assessment
4. Interpretation

3.1 Goal and Scope Definition

The importance of the Goal and scope definition—G&SD cannot be overestimated. Of course, the goal of an LCA study has to be presented at the beginning of any study as an introduction and gives some information about the background, etc. The far greater role of G&SD in LCA (ISO 2006a, b) rests in the standards which are strict in some items (structure, origin of data, reporting, reviewing) but very loose and open in others, even important ones. In LCIA, for instance, no impact categories, indicators and characterisation factors are prescribed, not even a recommended default list is given. But any LCA study has to include a (well-founded) list of impact categories, one is (for good reasons) not enough, see the 'footprint' discussion in Sect. 2.4.

On the other hand, the standards contain many "do not" orders out of the old—and not unfounded—fear of misuse. G&SD defines which rules—under the umbrella of the standards—are applied for a specific study. It is interesting to note that some other standards referencing ISO 14040 as the base standard require a collective G&SD for a group of related products in the form of so-called product category rules (PCR). This is the case for environmental product declaration (EPD), also called 'Level 3 labelling' in standard 14025 (ISO 2006c).

Standards are conventions, not laws, although many standards become *de facto* laws, especially in ordinances which explain how a law has to be used in practice. With regard to LCA, methods that deviate from the ISO standards or national standards related to the ISO norms can, of course, be used. It would be unlawful, however, to claim that a specific study has been performed according to ISO and actually was not. This problem is related to the need of critical reviews to be discussed again in Sect. 3.4. The type of critical review (if there is any) should be declared in G&SD including the names of the reviewers, if already known.

3.2 Life Cycle Inventory Analysis

Life Cycle Inventory Analysis—LCI forms the 'core' of any LCA study. It is also considered to be the most quantitative and scientific component (but see Sect. 7). The following steps are a minimum requirement to be fulfilled in comparative LCAs for all systems studied:

- System definition(s) including graphical presentation of the product trees
- Definition of the functional unit and the reference flow(s)[13]
- Data collection (input and output)
- Implementation of the data into the system, applying a predetermined cut-off rule and allocation rules, if appropriate
- Performing the calculations

The first two items refer to the individual study to be performed, either as a stand-alone LCA or—more frequently and interesting—in the form of a comparative LCA. In most cases the practitioners will already have produced rough schemes during the G&SD or even before (e.g. writing a research proposal). Frequently, any commercial or home-made software is used to help in this phase. The definition of the functional unit—needs thinking and cannot be delegated to the computer. What is the main function of the system(s) to be studied and compared? Which details (e.g. purely aesthetic ones) are unlikely to influence the final result and can be neglected in defining the fu? Such are the questions to be answered. Chemical, physical and biological/(eco-)toxicological knowledge is necessary in this phase and, of course, information about the production and waste management/recycling technology has to be collected in a systematic way. Multi-use systems need reliable data about average trip-numbers (e.g. in the case of refillable bottles). Questions about the lifetime of products and duration of services have to be answered.

The third item (data) is split, in most cases, into foreground data (to be supplied by the producer) and background data (mostly provided by generic data, e.g. from public and private data banks). It is essential that the data are representative for the reference-year or -period and for the region in which the processes occur. Rules for establishing data collections have recently been published (Sonnemann and Vigon 2011). Foreground data should be requested from producers (e.g. by using data collection sheets, for example see ISO 2006, annex A) and may be made unidentifiable

[13] The 'reference flow' is the translation of the verbally defined functional unit into technical terms.

for reasons of confidentiality (at least three producers are necessary). In this working phase it should already be known which data are essential for the LCIA stage. Possible data asymmetries should already be known or removed in a following update/correction step (see the double arrow discussion above in Sect. 2.3). A data quality discussion is indispensable.

The last item can be performed by excel-sheet calculations, using commercial software or applying the matrix method (Heijungs and Suh 2002, 2006) which can deal with loops in the product tree. In a recent development of the last method, even LCC can be combined with LCI paving the way to the calculation of LCSA (Heijungs et al. 2012).

The main result of LCI is the inventory table listing all inputs and outputs per unit process and aggregated per fu. For partial LCIs (e.g. cradle-to-factory gate or cradle-to-point of sale instead of cradle-to-grave) in the case of a tangible product (good) the results can also be related to a mass unit (mostly per kg or t = 1000 kg).

3.3 Life Cycle Impact Assessment

The somewhat loose content of Life Cycle Impact Assessment—LCIA mentioned in Sect. 3.1 is not—or not only—the result of the openness of the authors of ISO 14042 (ISO 2000a), the first international standard on LCIA. It was also caused by the fact that the European delegates, advised by SETAC Europe and its LCA steering committee (Udo de Haes et al. 1999a, b, 2002) and the US delegates, also advised by a SETAC (US) working group (Barnthouse et al. 1998) had substantially different ideas about impact assessment in LCA. Essentially the Europeans relied more on the precautionary principle, 'less is better' and 'beyond compliance' (Udo de Haes et al. 1999a, b), whereas the Americans favoured risk assessment and compliance with existing legislation (Barnthouse et al. 1998). The solomonic compromise consisted in accepting the European form, but not the recommended list of impact categories (essentially Heijungs et al. 1993).

The form of LCIA is given by the revised ISO 14040 (ISO 2006a) in a list of mandatory and optional elements:

A—Mandatory elements

- Selection of impact categories, category indicators and characterisation models
- Assignment of LCI results (classification)
- Calculation of category indicator results (characterisation)
- **Category indicator results** (LCIA results, LCIA profile)

B—Optional elements

- Calculation of the magnitude of category indicator results relative to reference information (normalisation)
- Grouping
- Weighting

The first mandatory element should already exist in G&SD, but has to be refined now. Also the LCI results should have been produced in the second stage with regard to the impact categories selected (classified), again final decisions to be taken now. The characterization is a genuine impact assessment element and requires knowledge of the interrelation between releases into the environment or extractions out of the environment and the potential impacts of the releases and extractions. In praxis, however, well established methods are selected by the practitioners, many being available in the software. The methods should be accepted internationally.

Among the optional elements 'normalization' is most frequently used. The results of this element allow discarding impacts contributing only marginally to the environmental damage in the region concerned.

The element 'weighting' *shall not be used in LCA studies intended to be used in comparative assertions intended to be disclosed to the public* (ISO 2006b § 4.4.5, Finkbeiner et al. 2006) (see also 'critical review' in Sect. 3.4). In contrast to 'grouping' according to (qualitative) similarities in the pattern of results (UBA 1999), weighting aims at quantitative aggregation of results using weighting factors. The result may be a single number which could be misused, hence the verdict by ISO.

3.4 Interpretation

Interpretation replaced older concepts such as improvement assessment (Sect. 2.2) and/or valuation (much discussed in Germany, see Klöpffer and Grahl 2009, 2014). The reason for declining the latter by ISO was that 'subjective' elements should be avoided in the otherwise scientific (as far as possible objective, see Sect. 6) LCA. Although "subjective is not (necessarily) arbitrary" (Klöpffer 1998b), since international conventions have another weight than opinions held by individuals, the last stage of LCA had to be profoundly changed (ISO 2000b; Lecouls 1999). Interpretation is now the counterpart of G&SD and essentially has to secure that the first three phases of an LCA study are well tuned in and consistent with each other. In addition, the plausibility and accuracy of the results have to be checked with suitable methods, such as sensitivity analyses and/or error calculations.

A critical review by independent experts should be performed for each LCA (optional) (ISO 14044, Sect. 6.2) but has to be done if the study is "intended to be used in comparative assertions intended to be disclosed to the public". In that case, the review has to be performed in the strongest form according to the panel method (ISO 14044, Sect. 6.3).

The structure of the final stage Interpretation is described in ISO 14044, Sect. 4.5 as follows:

- Identification of significant issues
- Evaluation with the elements Completeness check, Sensitivity check and Consistency check
- Conclusions, limitations and recommendations

Reporting (ISO 14044 Sect. 5) and Critical review report (ISO 14044 Sect. 6) are separate items outside of interpretation, but evidently belong to the final phase of any LCA study. The performance of the critical review should preferably be done in the interactive or accompanying mode, as proposed by (SETAC 1993). According to ISO 14040+44, it can also be performed 'a posteriori' in most cases, if the draft final report is available (Klöpffer 2012).

4 The Structure of LCA Beyond ISO 14040

A structure beyond the standard which defines the method seems at first sight to be an oxymoron. For this work, however, we need some structure for the elements and methods not (yet) or not fully covered by ISO 14040 ff. The following considerations may serve as guidance to such a structure.

4.1 Applications of Life Cycle Assessment

The applications of LCA lie outside the framework of LCA, see Fig. 2.3. The applications named explicitly in DIN 14040 are:

- Product development and improvement
- Strategic planning
- Public policy making
- Marketing

Actual LCA studies on these and other important topics certainly belong to this field of applied research. In addition to industry with an obvious interest in environmental assessment, also governments and related bodies (e.g. the European Commission) use increasingly LCA for policy making.

Marketing with environmental arguments and claims, based on environmental labels and product declarations (EPDs), is also regulated by the international standard series ISO 14020 ff. The most demanding standard 14025 (ISO 2006c) is firmly based on LCA (ISO 2006a, b) and product category rules (PCR), a kind of common G&SD for groups of similar products.

4.2 Beyond the Classical ISO LCA

In recent years, several methodological developments took place which go beyond the classical LCA, but are considered by the majority of LCA scientists as developments within Life Cycle Assessment. Such developments are, e.g., the consequential LCA (Weidema 2000) and the Input/Output (I/O) LCA and the I/O hybrid LCA (Suh 2003).

Other developments, mostly designated as 'footprints' (Carbon Footprint, Water Footprint) specify one impact of LCIA (e.g. Carbon Footprint = Global Warming Potential—GWP).

Methods requiring a high degree of regionalisation pose proportionally increased demands on both LCI (system definition and data acquisition) and LCIA (regional or even site-specific impacts) (Potting and Hauschild 2006). The regionalisation inevitably needed in social LCA (SLCA) (UNEP/SETAC 2009) and, thus, Life cycle Sustainability Assessment—LCSA, (Valdivia et al. 2011), may lead to a re-appraisal of regionalisation in environmental LCA, too.

Future updates of the ISO 14040 ff may specify how these and other new developments should be integrated.

4.3 Life Cycle Management

Life Cycle Management—LCM is an important concept in industry (UNEP/SETAC 2007), based on Life Cycle Thinking (LCT), but the approach is broader and goes beyond standardised LCA. Typically, a so-called 'tool box' is used whose composition depends on the nature of the business, on the size of the enterprise, its products, markets and the goals to be achieved—internally and externally. Simplified LC methods, such as 'MIPS'[14] (Schmidt-Bleek 1994), 'CED'[15] (VDI 1997) or 'EPS'[16] (Steen and Ryding 1992) can be used in addition to elaborate ones. Also methods developed in the time of the proto-LCAs are still in use, albeit in improved form, e.g. the Swiss eco-point method (Ahbe et al. 1990; Frischknecht et al. 2009).

An 'eco-efficiency' method developed and much used by the global largest chemical company BASF is based on LCA in the environmental part (Saling et al. 2002), but has to be combined with LCC in order to assess the ecological and economic aspects of product systems (Kicherer et al. 2007). An international standard for eco-efficiency assessment of product systems has been developed recently (ISO 2012). Eco-efficiency assessment of product systems (LCA+LCC) is also considered by scientists as a feasible first step toward a quantitative Life cycle sustainability assessment (Heijungs et al. 2012).

4.4 Life Cycle Sustainability Assessment

Life Cycle Sustainability Assessment—LCSA has been introduced in Sect. 2.4 as a new trend in LC methods. It can and should be used in LCM, but cannot be quantified by means of an ill-defined 'tool box'. It also goes further than 'eco-efficiency', adding the social component SLCA to give this 'three pillar' (or 'triple bottom line')

[14] Material Intensity Per Service unit (service unit = functional unit).

[15] Cumulative Energy Demand.

[16] Environmental Priority System.

structure which is typical for sustainability assessment (Valdivia et al. 2011). Since the term 'sustainability' (derived from the German word 'Nachhaltigkeit'[17] and originally used in forestry) (Carlowitz 2000; Grober 2010) is used today mostly in a loose—not to say sloppy—way meaning anything from environmental friendly, green, low carbon or even constant economic growth rates (!), etc., it has to be defined narrower for the purpose of sustainable product assessment (Curran 2012). The product- or 'micro' level (CALCAS 2009) is the one best suited for LCA. The 'meso' level (industrial production sites, companies, etc.) are also accessible to LCA as long as a reasonable functional unit and system boundaries can be defined, but the 'macro' level (national and global economy) needs a more elaborate system analysis, including non linearity, feed-backs, loops, etc. The authors of CALCAS (2009) use the acronym LCSA for Life Cycle Sustainability Analysis (instead of Assessment). This seemingly obsolete term, reminding the time of the proto-LCAs (Sect. 2.1) may indicate that ISO LCA (with A=Assessment) cannot cope with the full breath of LC studies (micro-meso-macro). In this context it should be mentioned that also political papers (e.g. the outcome of the UNEP conference in Johannesburg 2002) use the term life cycle analysis; actually it was this conference which opened the way for LCA to the global stage.

Since the (environmental) LCA is well established and standardised, and the environment is the main concern for a long-term survival of humankind, it seems logical to start LCSA with ISO LCA and add LCC and SLCA in a compatible way (Sect. 2.4). Both new components are defined and described in the form of a code of practice or guideline (Swarr et al. 2011; UNEP/SETAC 2009), and also the overarching concept of LCSA got an initiation toward standardisation (Valdivia et al. 2011). There is a good chance that in the future international standards will follow.

4.5 LCA Worldwide

A last development should be highlighted: LCA and the related methods are going global. Certainly, the UNEP/SETAC Life Cycle Initiative was the main driving force, but the UNEP world conferences in Rio de Janeiro (1992), Johannesburg (2002) and again in Rio de Janeiro (2012) paved the way for the world-wide dissipation and acceptance of life cycle thinking, before the detailed developments started. In Rio (1992), sustainable development was declared as the guiding principle for the twenty-first century. During the Johannesburg conference Life cycle thinking and LC methods were specifically introduced and recommended. It is certainly not by coincidence that in the same year 2002, in Prague, the UNEP/SETAC Life Cycle Initiative has been founded (Töpfer 2002).

Meanwhile, LCA societies, networks and individual working groups have been founded in many regions. These very welcome 'newcomers' add up to practitioners and researchers in the regions where LCA was first developed (USA/Canada, Europe, Japan/South Korea, Australia). The new regions bring environmental prob-

[17] as adjective: nachhaltig (German) → soutenu (French) → sustainable (English) (Grober 2010).

lems not in the focus of the 'old' LCA regions, as water scarcity, desertification, irreversible destruction of the natural forests, not to speak of the social problems to be dealt with in SLCA.

Regionalisation brings enormous data problems and it is hoped that the new regions start in the same way as the 'old' ones did: with the creation of national/regional data bases. Suitable impact assessment methods for the specific problems mentioned have to be developed, too. Then a true 'World LCA' may just appear at the horizon.

5 Structure of 'LCA Compendium'

5.1 Background and Future Prospects in Life Cycle Assessment

In this introductory volume, the authors present the basic features of LCA and how it came about under different perspectives. This first volume of the work may serve as stand-alone reading for the non-specialist (e.g. managers or policy makers, lay people interested in environmental, life cycle and sustainability assessment) wishing introductory but nevertheless in-depth information on the whole field.

The following volumes to come (although presumably not in this order) are align with the LCA structure defined by ISO (Sect. 3 and this volume, Chap. 3[18]). They are of prime importance for researchers trying to improve the method(s) as well as for practitioners using them in real-life LCA studies:

- Goal and Scope Definition in Life Cycle Assessment
- Life Cycle Inventory Analysis
- Life Cycle Impact Assessment
- Interpretation, Critical Review and Reporting

The second category of volumes goes beyond the limits set by ISO (Sect. 4 and this volume, Chap. 4[19]) and includes new developments, fields of LCA application, special types of LCA, Life Cycle Management—LCM and Life Cycle Sustainability Assessment—LCSA.

5.2 Goal and Scope Definition in Life Cycle Assessment

Goal and scope definition (G&SD, Sect. 3.1) in LCA is much more than a kind of introduction or exposé. It constitutes a vital part of any LCA and defines which rules

[18] The international standards as the constitution of LCA: the ISO 14040 series and its offspring by Matthias Finkbeiner.

[19] The UNEP/SETAC Life Cycle Initiative by Guido Sonnemann and Sonia Valdivia.

are valid for a specific LCA study within the frame of the international standards ISO 14040 and 14044 (ISO 2006a, 2006b).

The following topical areas are anticipated to be discussed in this volume. Since titles may change as work progresses, readers are encouraged to refer to the volume for the exact listing:

Envisioning the goal statement

- Determining the goal of an LCA application
- Private sector
  - process improvement
  - product design and development
  - supply chain management
- Public sector
  - public policy support
  - product category rules in Environmental Product Declarations (EPDS)
  - eco-labelling

Developing the G&SD

- Writing a clear goal statement and functional unit
- Defining the scope and study boundaries
- Matching the data needs to the goal
  - streamlined LCA
  - full mode LCA with complete transparency
- Identifying impact categories

Matching the goal to methodology

- Attributional LCA methodology
- Consequential LCA methodology
- Decisional LCA methodology
- Iterations in LCA practice: revisiting the goal
- Connection to the interpretation phase

5.3 Life Cycle Inventory Analysis

The second phase of Life Cycle Assessment—Life Cycle Inventory Analysis (LCI)—relates to system modelling methods and data acquisition. LCI is the oldest phase of LCA and is considered to be the most scientific one. Despite the long tradition, accurate descriptions of the actual procedure are rare (Heijungs et al. 2013). Inventory analysis means to build a system model according to the requirements of the goal & scope definition.

The following topical areas are anticipated to be discussed in this volume:

1. Introduction
   - Purpose and structure of LCI

   - History of LCI
   - Harmonisation and consensus building (SETAC, ISO, UNEP/SETAC)
   - LCI according to ISO standards
   - Scope dependency of LCI
   - Literature of LCI fundamentals (methods and data)
2. System delimitation and system boundaries
3. Development of unit process datasets
4. Special issues related to modelling and data
   - Time horizon (long-term emissions)
   - Biogenic CO_2
   - Agricultural soil (technosphere—biosphere boundary)
   - Accidents and incidents
   - Certificates and their role in LCI
   - Financial services, insurances, taxes, investments and the like
   - Transport services
   - Use phase
5. Allocation and recycling
6. Life cycle inventory data and databases
7. Data quality descriptors and indicators
8. The algorithms of life cycle inventory analysis
9. Aggregated inventory indicators
10. Visualisation tools
11. Links with LCIA
12. Perspectives and developments in LCI

5.4 *Life Cycle Impact Assessment*

Life Cycle Impact Assessment—LCIA (Sect. 3.3) is a permanently evolving phase of LCA consisting of about 10–15 well developed impact categories (IC) and new ones which are still in development. The degree of regionalisation has strong repercussions to the LCI phase, especially with regard to region-specific data (e.g. temperature of air and water; soil-dependent parameters).

The following topical areas are anticipated to be discussed in this volume:

Introduction

- Purpose and structure of LCIA
- History of LCIA
- Harmonisation and consensus building (SETAC, ISO, UNEP/SETAC)
- LCIA according to the ISO standards
- Midpoint and endpoint modelling (complementary approaches, strengths and weaknesses, areas of protection, definition and modelling)

Selection and classification

- Purpose of selection and classification

- Choice of impact categories (ICs) at midpoint
- Assignment of emissions to ICs (parallel and serial)

Characterisation modelling for individual Impact Categories (ICs)
General structure to be applied for each impact category:

- Principles and fundamentals of characterisation modelling
- Characteristics of the impact

It is planned to include the following ICs as individual chapters

- Climate change
- Stratospheric ozone depletion
- Human toxicity (carcinogens, non-carcinogens)
- Particulates
- Ionising radiation
- Photochemical oxidants (human health and vegetation impacts)
- Eco-toxicity
- Acidification
- Eutrophication/nutrient enrichment (freshwater, marine, terrestrial
- Land use (including salination, erosion and soil loss)
- Water use
- Abiotic resource use

Normalisation

- Purpose of normalisation
- Principles and fundamentals of normalisation modelling
- Examples of Normalisation references from different LCIA methods

Valuation

- Purpose of valuation
- Grouping, ranking, weighting
- ISO on valuation and comparative assertions disclosed to the public
- Science-based and value-based elements, damage modelling as a take on the former
- Weighting methods
- Examples of weighting factors from different LCIA methods

5.5 Interpretation; and, Critical Review and Reporting

The title of this volume is based on the fourth phase of LCA according to ISO (Sect. 3.4) and extended by the important topics of review and reporting.

The term **life cycle interpretation** is defined in the ISO 14040 standards, the … "phase of life cycle assessment in which the findings of either the inventory analysis or the impact assessment, or both, are combined consistent with the defined goal and scope in order to reach conclusions and recommendations".

The term **review** means to conduct a critical review that is a way to deal with quality assurance under confidentiality agreement. If parts of the underlying data of an LCA study are confidential, but there is a need for credibility and transparency (always if comparative LCA studies intended to be disclosed to the public), a reviewer who has seen the full report, including the confidential data, may certify its validity. ISO also poses high standards on reporting, especially in the case of third party reports, i.e. for the external use of LCA studies.

The following areas are anticipated to be discussed in this volume:

Interpretation

- The genesis of interpretation
- Scientific outline of interpretation
- Interpretation: something different to assessment and valuation
- Benefits from the contribution analysis
- Differences between sensitivity analysis und scenarios
- Importance of interpretation in the case of offsettings beyond system boundaries
- How to deal with uncertainties
- Data quality analysis as part of Interpretation
- The context of results: how to draw clear conclusions

Critical Review

- Criteria of critical review
- Critical review versus verification: similarities and differences
- The review panel: how to provide an optimum of expertise
- Benefits from Critical Review under inclusion of interested parties
- Cost-benefit-analysis of critical review: a case study

Reporting

5.6 *Overview on LCA Applications*

LCA, initially mostly used for packaging systems (still an important field of application), has found many applications, especially since the method has been standardised. Despite the standardisation, each field of application needs special techniques—not to speak about the data—to be specified in G&SD (Goal and Scope Definition) of the individual studies or in group-generic G&SD in form of PCR (Product Category Rules).

The contributions to this volume will introduce the main fields of application.

The following topical areas are anticipated to be discussed in this volume:

1. LCA for agriculture
2. LCA for aquaculture
3. LCA for food production and consumption
4. Industrial products

5. Land use in LCA
6. Water use in LCA
7. Renewable energy systems
8. Bioenergy
9. Conventional energy systems
10. Wood and other renewable resources
11. Photovoltaics as an example where LCA and product development influenced each other
12. Packaging systems including recycling
13. LCA of waste management systems
14. Building products and whole buildings
15. Automotive industry
16. Policies in relation to LCA

5.7 Special types of Life Cycle Assessment

Variations of LCA were developed which are compatible with the SETAC-ISO LCA but not necessarily with all detailed prescription in ISO 14040+44 (ISO 2006a, b) (Sect. 4.2). Some of these methods are being standardised, but refer to ISO 14040 as reference standard (this volume, Chap. 3[20]).

The following topical areas are anticipated to be discussed in this volume:

- Carbon footprint
- Water footprint
- Eco-efficiency and resource efficiency assessment
- Input-output and hybrid LCA
- Material flow analysis
- Consequential LCA
- LCA of organizations

5.8 Life Cycle Management

Life Cycle Management—LCM embraces many applications of Life Cycle Thinking, product- as well as company-related LCAs and simplified methods not or not fully compliant with the ISO standards (Sect. 4.3). The methods used may also go beyond (environmental) LCA and contain Life Cycle Costing—LCC and Social Life Cycle Assessment—SLCA.

The following areas are anticipated to be discussed in this volume:

---

[20] The international standards as the constitution of LCA: the ISO 14040 series and its offspring by Matthias Finkbeiner.

What is life cycle management (LCM)?

- This section will include a synthesis of modern management systems

Sustainability and life cycle thinking

- The triple bottom line—the business case of sustainability
- Introduction to life cycle thinking for products
- Responsibilities in the life cycle and the value chain

Helpful tools and concepts for life cycle management

- Analytical life cycle approaches supported by data information and models
- Cleaner production
- Eco-design/ Design for environment (DfE)
- Dematerialization and industrial ecology
- Environmental performance and life cycle costing (LCC)
- Social life cycle assessment (SLCA), sustainability indicators and life cycle sustainability assessment (LCSA)
- Social responsibility, sustainability reporting and stakeholder engagement along the value chain—going beyond the organization boundaries
- Communication at the product level through declarations, labeling and certification
- Sustainable procurement and supply chain management
- Resource efficiency and eco-efficiency as well as integrated product policy (IPP) and sustainable consumption and production: concepts for action by the public and private sectors

Implementation of life cycle management

- From environmental management systems to product-orientated management systems
- Life cycle management in different departments
- Integration of concepts and tools for decision-making
- How life cycle approaches can be integrated into standard business processes
- Implementation of life cycle management—a step by step approach
- Policies and commitment working with cross functional teams
- Plan—Overview of status and set objectives and targets
- Do—Put the plan into practice and document the efforts and results
- Check—Evaluate
- Act—Revise and take it to the next level

LCM capability maturity model

LCM in practice

5.9 *Life Cycle Sustainability Assessment*

Sustainability was adopted by UNEP in Rio de Janeiro (1992) as the main political goal for the future development of humankind. It should also be the ultimate aim of product development. According to the well known interpretation of the original definition given in the Brundtland report, sustainability comprises three components: environment, economy and social aspects. These components or 'pillars' of sustainability have to be properly assessed and balanced if a new product is to be designed or an existing one is to be improved (Sect. 4.4).

The following areas are anticipated to be discussed in this volume:

History

- Scientific paradigm for sustainability: sustainability science
- LCA: divergence in developments (decade of elaboration), towards LCSA
- Lessons learnt from LCA
- LCA and sustainability

Life cycle sustainability assessment

- Life cycle sustainability assessment
- Life cycle sustainability analysis

From LCA to LCSA: broadening

- Environmental indicators
- Life cycle-based approaches taking into account social aspects
- Life cycle-based approaches taking into account economic aspects
- Weighting
- Interpretation and visualisation of results: approaches to integration of sustainability indicators
- Multi-scale analysis

From LCA to LCSA: deepening

- Modelling techniques for managing complexity in LCSA
- Including mechanisms into LCSA: option for modelling

Future research in LCSA

- How to further develop LCSA in order to be in line with the ontology and epistemology of sustainability science
- Framing the question
- Sustainability indicators for LCSA
- Mechanisms in empirical modelling for LCSA
- Cross-cutting research for integration
- Research and education
- Exemplary projects

5.10 *LCA Worldwide*

In addition to the regions in which LCA has been developed, mainly Europe, the USA, Canada, Japan and South Korea, several regions and continents have strongly increased their interest and capabilities in performing LCAs, often shown by LCA societies and regular conferences (Sect. 4.5). Such regions are Latin America, Australia and New Zealand, Continental East Asia and India. In addition, there are pioneer groups and individual LCA researcher in Africa. The UNEP/SETAC Life Cycle Initiative has strongly supported these activities since its foundation (Töpfer 2002; Valdivia et al. 2011).

6 New Developments and Special Types of Life Cycle Assessment—How Are they taken into Account?

Any living science and related fields of knowledge develop with time. Nevertheless, the editors believe that the field of LCA has matured enough to venture into this major editorial project. Since this task will take many years to finish, it will not be just a snapshot in time. Many aspects of LCA turned out to be quite stable, especially due to the secure frame provided by the standardisation (this volume, Chap. 3[21]) and also by tradition. It is even disappointing to see that many developments in the academic area (PhD dissertations and scientific publications) only slowly enter the daily routine of LCA practitioners. Users in industry and practitioners wish robust and simple-to-use (not necessarily simplistic) methods (Baitz et al. 2012). This is understandable and constitutes at the same time a request to academia to formulate new methods in such a way that they can be used in praxis.

We hope that this work will improve the dialogue between the 'Two planets', as we once called the different spheres which have difficulties in communication (this volume, Chap. 5; Klöpffer and Heinrich 2001). At that time, we thought that the journal alone will do the job, but in the meantime we have noticed that a more structured and in-depth publication is needed.

7 How Scientific is LCA?

This question is not easy to answer, although most practitioners agree that LCA is **science-based**, with the two inner phases (LCI and LCIA) more so than the outer ones (G&SD and Interpretation). Nevertheless, all four belong together (Sect. 3 and this volume, Chap. 3) for the good reasons discussed.

[21] The international standards as the constitution of LCA: the ISO 14040 series and its offspring by Matthias Finkbeiner.

At first glance, Life Cycle Inventory Analysis is the most scientific stage of LCA. Being essentially a material- and energy flow analysis, it is based on several of the best founded laws of natural science:

- The law of mass conservation
- The law of energy conservation (1st law of thermodynamics)
- The entropy law (2nd law of thermodynamics)
- Stoichiometry (one of the basic concepts of chemistry)
- Einstein's $E = mc^2$ (for nuclear reactions)

The precision of an inventory is therefore—theoretically—only depending on the quality of input data and the system boundaries chosen. This is only true, however, if no allocation (for co-products, in open-loop recycling, etc.) is needed in modelling. Allocation cannot be treated in a strictly scientific manner and the rules given by the ISO standards prefer the avoidance of allocation rather than proposing a convention which may come near to a truly scientific solution. Allocation is not the only flaw in LCI, but the most evident one. Inaccuracies in data and model assumptions can be treated by uncertainty calculations (Ciroth et al. 2004).

Life Cycle Impact Assessment (LCIA) seems to be less rigorous than LCI since the laws of biology and especially those dealing with eco-systems seem to be less 'hard' than those of physics and chemistry. In some important impact categories, however, physical, chemical, atmospheric sciences, water and soil science enter the characterisation of the impacts; and biology itself has a 'hard core' we can rely upon. Conventions are needed here, too, and ISO 14040 ff gives guidance how to proceed correctly.

At this point we should consider the role of conventions in borderline-cases of scientific truth. In general, the importance of conventions (including, of course, the standards) increases, if one or the other step in a scientific method cannot be fully met. One of the best examples is the international standard for units, the 'système international' (SI), the modern version of the old meter convention[22] (ISO 1981). In Life Cycle Assessment we have the series of standards 14040 ff defining LCA (this volume, Chap. 3, Finkbeiner 2012). Subjective elements are recommended only as a last resort and have to be secured by sensitivity calculations. Subjective is not necessarily arbitrary, however (Klöpffer 1998), since in connection with science the term 'subjective' only means not 'objective' (as, e.g., measurements in all sciences should be). In social sciences this high degree of objectivity is hardly to achieve and conventions become important. Designating all conventions as 'subjective' would imply that even global conventions, e.g. the human rights convention, have the same status as a purely subjective opinion by an individual human being. This is clearly nonsense and a closer look at the definition of subjectivity would be worthwhile.

An epistemological view on LCA would also be worthwhile. What would Karl Popper (Popper 1934, 1959) say about LCA? Can LCA results be falsified? The 'critical review by interested parties', as prescribed for certain LCA studies, may be

[22] First conceived during the French revolution.

considered as a kind of falsification procedure[23]. But even if LCA in general will not pass the very strict falsification test derived for the natural sciences, it will easily meet the requirements of the 'social sciences and humanities'.

LCA is multidisciplinary with scientists, scholars and practitioners coming from many different disciplines. All these disciplines have somewhat different concepts about defining and finding scientific truth. Common to all forms of modern science, however, is the publishing of all essential methods and results; this is also true for LCA (Klöpffer 2007). The ISO standards recommend a critical review for all LCA studies, but firmly prescribe the most rigorous form if comparative assertions are intended to be disclosed to the public. Clearly, this is primarily a measure against fraudulent manipulations of LCAs, but increases also the scientific quality of such studies. Publication in a scientific journal requires an additional peer review. LCA studies not disclosed to the public are also subject to reporting rules defined by ISO 14044.

[23] Birgit Grahl, private communication 2011.

Appendix—Glossary

Cradle-to-grave analysis	All important steps in the life cycle of a product are included in the analysis (extraction of raw materials from the environment (soil, water, air), the production of materials, the final products, their use and waste removal or recycling).
Downstream process	Toward the 'grave'
Footprint studies	Footprint studies are no full LCAs (see simplified/ streamlined LCA). They mostly contain only one impact category: 'Carbon footprint'=Global Warming Potential. The 'water footprint' can be integrated into LCIA as an impact category belonging to resource use.
Functional unit	The basis of comparison of product systems (goods *and* services) if they provide the same or a very similar function.
Life cycle sustainability assessment—LCSA	In life cycle product assessment, LCA deals with the environmental aspects only. In order to give the full picture, however, an economic and a social life cycle assessment have to be added to the environmental one. In the 'three pillar' interpretation of sustainability, environmental, economic and social aspects have to be considered and weighted against each other. LCSA=LCA+LCC+SLCA LCSA: Life Cycle Sustainability Assessment LCA: (environmental) Life Cycle Assessment LCC: (environmental) Life Cycle Costing SLCA: Social Life Cycle Assessment
Product tree	The most common form of graphical presentation of product life cycles.
proto-LCAs	The early LCAs before harmonisation (SETAC) and standardisation (ISO).
SETAC triangle	Model of the phases (components) of an LCA. The first SETAC triangle was developed at the SETAC workshop 'A Technical Framework for Life Cycle Assessments' in August 1990, Smugglers Notch, Vermont and consisted of three components: Inventory, Impact Analysis, Improvement Analysis. The SETAC triangle 1992 (Sandestin workshop) and 1993 (Sesimbra workshop) consisted of four components: Goal Definition and Scoping, Inventory Analysis, Impact Assessment and Improvement Assessment. In the course of the ISO standardization process 'Improvement assessment' was replaced by 'Interpretation'.
Simplified LCA/ Streamlined LCA	In 1994 the LCA Steering Committee of SETAC Europe established the Workgroup Screening and Streamlining. In the same year, the SETAC North America workgroup on Streamlining LCA was initiated. Both groups concluded their multi-year efforts on the issue of Simplifying/Streamlining by a report in each case. The approaches of the reports are different. 1. The report of SETAC Europe discusses the methods for producing simplified procedures, commonly described as screening LCA studies, streamlined LCA studies and simplified LCA studies. 2. The report of SETAC North America is more a description of carefully planning and stating an LCA's goal than it is about Streamlined LCA methodology.

Simplified LCA	Simplified LCA is an application of the LCA methodology for a comprehensive screening assessment. A simplified LCA should cover three steps which are iteratively interlinked: 1. *Screening:* identifying those parts of the system (life cycle) or of the elementary flows that are either important or have data gaps. 2. *Simplifying:* using the findings of Screening in order to focus further work on the important parts of the system or of the elementary flows. 3. *Assessing reliability:* checking that simplifying does not significantly reduce the reliability of the overall result. Simplifying methods can reduce the complexity of an LCA and so reduce the cost, time and effort required, by exclusion of certain life cycle stages, system inputs or outputs or impact categories, or use of generic data modules for the system under study.
Streamlined LCA	Identification of elements of an LCA that can be omitted or where surrogate or generic data can be used without significantly affecting the accuracy of the results. Streamlining LCA is a practice to make a detailed/full LCA more manageable. Streamlining LCA can be achieved in a number of ways, including: – Limiting the scope in terms of time, cost, data, analytical approach: for example, eliminating life cycle phases deemed not significant, or processes with negligible effect on the environment; – Use of qualitative information; – Removal of upstream and/or downstream components; – Use of specific impact category.
Supply chain	Usual, but misleading (since suggesting linearity) designation of the upper part of a product tree or branches thereof.
System boundary	The system boundary separates the system to be studied from the rest of the technosphere and the environment.
Unit process	The smallest unit for which data are available.
Upstream process	Toward the 'cradle'.
Use phase	The use phase is the centre of most life cycles defined in LCA.

References

Ahbe S, Braunschweig A, Müller-Wenk R (1990) Methodik für Oekobilanzen auf der Basis ökologischer Optimierung. In: Bern (ed) Bundesamt für Umwelt, Wald und Landschaft (BUWAL). Schriftenreihe Umwelt no 133

Baitz M, Brauner E, Albrecht S, Broadbent C, Castellan G, Conrath P, Fava J, Finkbeiner M, Fischer M, Fullana i Palmer P, Krinke S, Leroy C, Loebel O, McKeown P, Mersiowsky I, Möginger B, Pfaadt M, Rebitzer G, Rother E, Ruhland K, Schanssema A, Tikana L (2013) LCA's theory and practice: like ebony and ivory living in perfect harmony? Int J Life Cycle Ass 18(1):5–13

Barnthouse L, Fava J, Humphreys K, Hunt R, Laibson L, Noesen S, Norris G, Owens J, Todd J, Vigon B, Weitz K, Young J (eds) (1988) Life-cycle impact assessment: the state-of-the-art. Report of the SETAC life-cycle assessment (LCA) impact assessment workgroup, 2nd edn. Society of Environmental Toxicology and Chemistry, Pensacola

Benoît C, Norris GA, Valdivia S, Ciroth A, Moberg A, Bos U, Prakash S, Ugaya C, Beck T (2010) The guidelines for social life cycle assessment of products: just in time! Int J Life Cycle Ass 15(2):156–163

Berger M, Finkbeiner M (2010) Water footprinting: how to address water use in life cycle assessment? Sustainability 2:919–944. doi: 10.3390/su2040919

Blouet A, Rivoire E (1995) L'Écobilan. Les produits et leurs impacts sur l`environnement. Dunod, Paris (ISBN 2–10-002126-5)

Boustead I (1993) Eco-profiles of the European plastics industry. Report 3: polyethylene and polypropylene. Report to the European centre for plastics in the environment (PWMI), Brussels

Boustead I (1996) LCA—how it came about. The beginning in the UK. Int J Life Cycle Ass 1(3):147–150

Boustead I, Hancock GF (1979) Handbook of industrial energy analysis. Ellis Horwood Ltd, Chichester (Wiley, New York)

BSI—British Standards Institution (ed) (2008) Publicly available specification (PAS) 2050:2008. Specification for the assessment of the life cycle greenhouse gas emissions of goods and services

BUS—Bundesamt für Umweltschutz (ed) (1984) Oekobilanzen von Packstoffen. Schriftenreihe Umweltschutz no 24, Bern

CALCAS, Zamagni A, Buttol P, Buonamici R, Masoni P, Guinée JB, Huppes G, Heijungs R, van der Voet E, Ekvall T, Rydberg T (2009) Co-ordination action for innovation in life-cycle analysis for sustainability. D20 Blue Paper on Life Cycle Sustainability Analysis. Revision 1 after the open consultation. http://www.estis.net

Carlowitz HC von (1713, 2000): Sylvicultura oeconomica. Anweisung zur wilden Baum-Zucht. Johann Friedrich Braun, Leipzig, (Reprint: TU Bergakademie Freiberg. ISBN 3-86012-115-4. Freiberg, Germany)

Christiansen K (1997) Simplifying LCA: just a cut? Final report of the SETAC-Europe LCA screening and streamlining working group. Brussels

Ciambrone DF (1997) Environmental life cycle analysis. Lewis Publishers, Boca Raton

Ciroth A, Fleischer G, Steinbach J (2004) Uncertainty calculation in life cycle assessments. A combined model of simulation and approximation. Int J Life Cycle Ass 9(4):216–226

Curran MA (ed) (1996) Environmental life-cycle assessment. McGraw-Hill, New York (ISBN 0-07-015063-X)

Curran MA (2008) Development of life cycle assessment methodology: a focus on co-product allocation. PhD Thesis, Erasmus University

Curran MA (ed) (2012, Nov) Life cycle assessment handbook: a guide for environmentally sustainable products. Wiley, 640 p (ISBN: 978-1-1180-9972-8; also available as e-book)

Curran MA, Young S (1996) Report from the EPA conference on streamlining LCA. Int J Life Cycle Ass 1(1):57–60

Ekvall T, Tillman A-M (1997) Open-loop recycling: criteria for allocation procedures. Int J Life Cycle Ass 2(3):155–162

Fava JA, Denison R, Jones B, Curran MA, Vigon B, Selke S, Barnum J (eds) (1991) SETAC workshop report: a technical framework for life cycle assessments. August 18–23 1990, Smugglers Notch, Vermont. SETAC, Washington, DC

Fink P (1997) LCA—how it came about. The roots of LCA in Switzerland: continuous learning by doing. Int J Life Cycle Ass 2(3):131–134

Finkbeiner M (2009) Carbon footprinting—opportunities and threats. Int J Life Cycle Ass 14(2):91–94

Finkbeiner M (2013) From the 40s to the 70s—the future of LCA in the 14000 family. Int J Life Cycle Ass 18(1):1–4

Finkbeiner M, Matsuno Y (eds) (2000a) Special issue Japan. Int J Life Cycle Assess 5(5):253–316

Finkbeiner M, Matsuno.Y (2000b) LCA in Japan—the past, the present, the future. Editorial Int J Life Cycle Ass 5(5):253–254

Finkbeiner M, Inaba A, Tan RBH, Christiansen K, Klüppel H-J (2006) The new international standards for life cycle assessment: ISO 14040 and ISO 14044. Int J Life Cycle Ass 11:80–85

Finkbeiner M, Schau EM, Lehmann A, Traverso M (2010) Towards life cycle sustainability assessment. Sustainability 2:3309–3322. doi:10.3390/su2103309

Franke M (1983) Umweltauswirkungen durch Getränkeverpackungen: Systematik zur Ermittlung der Umweltauswirkungen von Prozessen am Beispiel von Einweg- und Mehrweg-Getränkebehältern. Dissertation, Technical University Berlin

Frischknecht R, Steiner R, Jungbluth, N (2009) Methode der ökologischen Knappheit—Ökofaktoren 2006. Methode für die Wirkungsabschätzung in Ökobilanzen. Umwelt-Wissen Nr 0906. Bundesamt für Umwelt (BAFU) Bern

Gabathuler H (1997) The CML story. How environmental sciences entered the debate on LCA. Int J Life Cycle Ass 2(4):187–194

Grober U (2010) Die Entdeckung der Nachhaltigkeit. Kulturgeschichte eines Begriffs. Antje Kunstmann, München

Habersatter K, Widmer F (1991) Oekobilanzen von Packstoffen. Stand 1990. In: Bundesamt für Umwelt, Wald und Landschaft (BUWAL) (ed), Schriftenreihe Umwelt No 132, Bern

Hauschild M, Wenzel H (1998) Environmental assessment of products. scientific background, vol 2. Chapman & Hall, London, (ISBN 0-412-80810-2)

Heijungs R, Guinée JB, Huppes G, Lamkreijer RM, Udo de Haes HA, Wegener Sleeswijk A, Ansems AMM, Eggels PG, van Duin R, de Goede HP (1993) Environmental life cycle assessment of products. Guide (Part 1) and Backgrounds (Part 2). October 1992, prepared by CML, TNO and B & G. English Version Leiden

Heijungs R, Frischknecht R (1998) On the nature of the allocation problem. A special view on the nature of the allocation problem. Int J Life Cycle Ass 3(6):321–332

Heijungs R, Suh S (2002) The computational structure of life cycle assessment. Kluwer Academic, Dordrecht, (ISBN 1-4020-0672-1)

Heijungs R, Suh S (2006) Reformulation of matrix-based LCI: from product balance to process balance. J Clean Prod 14(1):47–51

Heijungs R, Settanni E, Guinée J (2013) Toward a computational structure for life cycle sustainability analysis: unifying LCA and LCC. Int J Life Cycle Ass 18(9):1722–1733, doi: 10.1007/s11367-012-0461-4

Heintz B, Baisnée P-F (1992) System boundaries. In: SETAC-Europe (ed) Life-cycle assessment. Workshop Report, 2–3 December, Leiden 1991. Brussels

Hochschorner E, Finnveden G (2003) Evaluation of two simplified life cycle assessment methods. Int J Life Cycle Ass 8(2):119–128

Hunkeler D (2006) Societal LCA methodology and case study. Int J Life Cycle Ass 11(7):371–382

Hunkeler D, Yasui I, Yamamoto R (1998) LCA in Japan: policy and progress. Int J Life Cycle Ass 3(3):124

Hunkeler D, Lichtenvort K, Rebitzer G (eds) (2008) Environmental life cycle costing. CRC Press, SETAC, Boca Raton

Hunt RG, Sellers JD, Franklin WE (1992) Resource and environmental profile analysis: a life cycle environmental assessment for products and procedures. Environ Impact Asses Rev 12:245–269

Hunt RG, Boguski TK, Weitz K, Sharma A (1998) Case studies examining streamlining techniques. Int J Life Cycle Ass 3(1):36–42

Hunt R, Franklin WE (1996) LCA—how it came about. Personal reflections on the origin and the development of LCA in the USA. Int J Life Cycle Ass 1(1):4–7

Huppes G, Schneider F (eds) (1994) Proceedings of the European workshop on allocation in LCA. Under the auspices of SETAC-Europe. Leiden, (ISBN 90-5191-078-9)

ISO (1981) International Standard Organization: SI units and recommendations for the use of their multiples and of certain other units, 2nd edn. ISO, Geneva, (1000:1981)

ISO (1997) International Standard Organization: environmental management—life cycle assessment—principles and framework. ISO, Geneva, (14040:1997)

ISO (1998) International Standard Organisation: environmental management—life cycle assessment—goal and scope definition and inventory analysis. ISO, Geneva, (14041:1998)

ISO 14042 (2000a) International Standard Organisation: environmental management—life cycle assessment—life cycle impact assessment. ISO, Geneva, (14042:2000)

ISO (2000b): International Standard Organisation: environmental management—life cycle assessment—interpretation. ISO, Geneva, (14043:2000)
ISO (2006a) International Standard Organization: environmental management—life cycle assessment—principles and framework. ISO, Geneva, (14040:2006)
ISO (2006b) International Standard Organisation: environmental management—life cycle assessment: requirements and guidelines. ISO, Geneva, (14044:2006)
ISO (2006c) International Standard Organisation: environmental management—environmental labels and declarations—Type III environmental declarations—principles and procedures ISO, Geneva, (14025)
ISO (2013) International Standard Organisation: carbon footprint of products—requirements and guidelines for quantification and communication ISO/DIS, Geneva, (14067)
ISO (2012) International Standard Organisation ISO/TC 207/SC5/WG7: environmental management—eco-efficiency assessment of product systems—principles, requirements and guidelines ISO, Geneva, (14045)
Jensen AA (2007) The LCM conferences—how they came about. Int J Life Cycle Ass 12 (Suppl 1):2–3
Jensen AA, Postlethwaite D (2008) SETAC Europe LCA steering committee—the early years. Int J Life Cycle Ass 13(1):1–6
Kicherer A, Schaltegger S, Tschochohei H, Ferreira Pozo B (2007).: Eco-efficiency. Combining life cycle assessment and life cycle costs via normalization. Int J Life Cycle Ass 12(7):537–543
Klöpffer W (1989) Von der Emotion zur Ökobilanz. (Balancing reality and emotional response in relevant environmental issues). The Rudolf Wild International Seminars 1989. Heidelberg
Klöpffer W (1998a) Ciambrone DF (1997) Environmental life cycle analysis. Book review Int J Life Cycle Ass 3(5):280, (ISBN 1-56670-214–3)
Klöpffer W (1998b) Subjective is not arbitrary. Editorial. Int J Life Cycle Ass 3(2):61–62
Klöpffer W (2003) Life-cycle based methods for sustainable product development. Int J Life Cycle Ass 8(3):157–159
Klöpffer W (2006) The role of SETAC in the development of LCA. Int J Life Cycle Ass 11 (Suppl 1):116–122
Klöpffer W (2007) Publishing scientific articles with special reference to LCA and related topics. Int J Life Cycle Ass 12(2):71–76
Klöpffer W (2008) Life-cycle based sustainability assessment of products. Int J Life Cycle Ass 13(2):89–94
Klöpffer W (2012) The critical review of life cycle assessment studies according to ISO 14040 and 14044: origin, purpose and practical performance. Int J Life Cycle Ass 17(9):1087–1093
Klöpffer W, Heinrich AB (2001) Two planets and one journal. Editorial. Int J Life Cycle Ass 6(1):1–3
Klöpffer W, Grahl B (2009) Ökobilanz (LCA)—Ein Leitfaden für Ausbildung und Beruf. Wiley-VCH, Weinheim
Klöpffer W, Grahl B (2014) Life cycle assessment (LCA)—a guide to best practice. Wiley-VCH, Weinheim
Lecouls H (1999) ISO 14043: environmental management—life cycle assessment—life cycle interpretation. Int J Life Cycle Ass 4(5):245
Lindfors L-G, Christiansen K, Hoffmann L, Virtanen Y, Juntilla V, Hanssen O-J, Rønning A, Ekvall T, Finnveden G (1995) Nordic guidelines on life-cycle assessment. Nordic Council of Ministers. Copenhagen, (Nord 1995:20)
Lundholm M P, Sundström G (1985) Ressourcen und Umweltbeeinflussung. Tetrabrik Aseptic Kartonpackungen sowie Pfandflaschen und Einwegflaschen aus Glas. Malmö
Marsmann M (1997) ISO 14040—the first project. Int J Life Cycle Ass 2(3):122–123
Marsmann M (2000) The ISO 14040 family. Int J Life Cycle Ass 5(6):317–318
Müller-Wenk R (1978) Die ökologische Buchhaltung. Ein Informations- und Steuerungsinstrument für umweltkonforme Unternehmenspolitik. Campus, Frankfurt am Main
Oberbacher B (1975) Ökologischer Nutzwert der Einwegflasche. Ein systemanalytischer Ansatz. Müll und Abfall Nr 8:234–238

Oberbacher B, Nikodem H, Klöpffer W (1996) LCA—how it came about. An early systems analysis of packaging for liquids which would be called an LCA today. Int J Life Cycle Ass 1(2):62–65

Popper KR (1934, 1959) Logik der Forschung. Springer, Wien 1934. 1st English edn: the logic of scientific discovery. Hutchison, London

Potting J, Hauschild MZ (2006) Spatial differentiation in life cycle impact assessment—a decade of method development to increase the environmental realism of LCIA. Int J Life Cycle Ass 11(Suppl 1):11–13

Remmen A, Jensen AA, Frydendal J (2007) The triple bottom line—the business case of sustainability. In: Life Cycle Management. A Business Guide to Sustainability. UNEP/SETAC Life Cycle Initiative. pp 10–11, (ISBN 978-92-807-2772-2)

Saling P, Kicherer A, Dittrich-Krämer B, Wittlinger R, Zombik W, Schmidt I, Schrott W, Schmidt S (2002) Eco-efficiency analysis by BASF: the method. Int J Life Cycle Ass 7(4):203–218

Schmidt-Bleek F (1994) Wieviel Umwelt braucht der Mensch? MIPS—Das Maß für ökologisches Wirtschaften. Birkhäuser Verlag, Berlin

SETAC-Europe (ed) (1992) Life-cycle assessment. workshop report, 2–3 December, Leiden 1991, Brussels

SETAC-Europe (2011) Annual meeting in Milano. Founding session of an ecological footprint working group.

SETAC—Society of Environmental Toxicology and Chemistry (1993) Guidelines for life-cycle assessment: a 'Code of Practice'. From the SETAC workshop held at Sesimbra, Portugal 31 March—3 April 1993. Brussels and Pensacola, Florida

Sinden G (2009) The contribution of PAS 2050 to the evolution of international greenhouse gas emission standards. Int J Life Cycle Ass 14(3):195–203

Sonnemann G, Vigon B (eds) (2011): Global guidance principles for life cycle assessment databases—'Shonan Guiding Principles'. 156 pp. ISBN: 978-92-807-3174-3. UNEP/SETAC Life Cycle Initiative, Paris. http://lcinitiative.unep.fr

Steen B, Ryding S-O (1992) The EPS enviro-accounting method. IVL Report, Göteborg 1

Suh S (2003) Input-output and hybrid life cycle assessment. Int J Life Cycle Ass 8(5):257

Swarr T, Hunkeler D, Klöpffer W, Pesonen H-L, Ciroth A, Brent A, Pagan B (2011) Environmental life cycle costing: a code of practice. SETAC Press, Pensacola

Töpfer K (2002) The launch of the UNEP-SETAC Life Cycle Initiative (Prague, April 28 2002). Int J Life Cycle Ass 7:191

UBA (1999) Schmitz S, Paulini I (eds) Bewertung in Ökobilanzen. Methode des Umweltbundesamtes zur Normierung von Wirkungsindikatoren, Ordnung (Rangbildung) von Wirkungskategorien und zur Auswertung nach ISO 14042 und 14043. Version `99. UBA Texte 92/99, Berlin

Udo de Haes HA, Jolliet O, Finnveden G, Hauschild M, Krewitt W, Müller-Wenk R (1999a) Best available practice regarding impact categories and category indicators in life cycle impact assessment. Part 1. Int J Life Cycle Ass 4(2):66–74

Udo de Haes HA, Jolliet O, Finnveden G, Hauschild M, Krewitt W, Müller-Wenk R (1999b) Best available practice regarding impact categories and category indicators in life cycle impact assessment. Part 2. Int J Life Cycle Ass 4(3):167–174

Udo de Haes HA, Finnveden G, Goedkoop M, Hauschild M, Hertwich EG, Hofstetter P, Jolliet O, Klöpffer W, Krewitt W, Lindeijer E, Müller-Wenk R, Olsen SI, Pennington DW, Potting J, Steen B (eds) (2002) Life-cycle impact assessment: striving towards best practice. SETAC Press, Pensacola, (ISBN 1-880611-54-6)

UNEP/SETAC (2007) Life cycle management. a business guide to sustainability. UNEP/SETAC Life Cycle Initiative. Paris, (ISBN 978-92-807-2772-2)

UNEP/SETAC (2009) Benoît C, Mazijn B (eds) Guidelines for social life cycle assessment of products. Paris. 104 p. http://lcinitiative.unep.fr

Valdivia S, Ugaya CML, Sonnemann G, Hildenbrand J (eds) (2011) Towards a life cycle sustainability assessment. Making informed choices on products. UNEP/SETAC Life Cycle Initiative, Paris, (ISBN: 978-92-807-3175-0)

Weidema B (2000) Avoiding co-product allocation in life-cycle assessment. J Ind Ecol 4(3):11–33

Weitz KA, Todd JA, Curran MA, Malkin MJ (1996) Streamlining life cycle assessment. Considerations and a report on the state of practice. Int J Life Cycle Ass 1(2):79–85

Wenzel H, Hauschild M, Alting L (1997) Environmental assessment of products, vol 1: methodology, tools and case studies in product development. Chapman & Hall, London, (ISBN 0-412-80800-5)

World Commission on Environment and Development (1987) Our common future (The Brundtland Report), Oxford

Wrisberg N, Udo de Haes HA, Triebswetter U, Eder P, Clift R (eds) (2002) Analytical tools for environmental design and management in a systems perspective. Kluwer Academic, Dordrecht

Chapter 2
The Role of the Society of Environmental Toxicology and Chemistry (SETAC) in Life Cycle Assessment (LCA) Development and Application

James A. Fava, Andrea Smerek, Almut B. Heinrich and Laura Morrison

Abstract Although there was a demand for environmental health data on chemicals, there was no global scientific organization able to talk about the science behind the regulations being developed. The Society of Environmental Toxicology and Chemistry (SETAC) was founded in 1979. SETAC has three strengths: its global scale, its tripartite membership and governance, and its scientific base. Because SETAC was developed on an international scale, it has been able to address global environmental issues.

The SETAC North American LCA Advisory Group is a formally recognized group within SETAC that has been in existence since June 1991. Similarly, SETAC Europe established an LCA Steering Committee. Both the LCA Advisory and Steering Committee are referred to as the SETAC LCA Groups.

The LCA Groups report to the Board of Directors of both SETAC and SETAC Europe. Specific activities such as workshops, conferences, or educational material development, including 'position papers', are approved by the Board of Directors. During the 1990s these SETAC LCA Groups were instrumental in driving the scientific progress to codify the professional practice of LCA. During this time period, several major workshops were successfully organized and over a dozen key publications produced. The SETAC LCA Groups also broadly supported the initial preparation of the ISO 14040 series of voluntary international standards as well as their subsequent revisions.

J. A. Fava (✉)
PE INTERNATIONAL, Inc., 626 Meadow Drive,
19380, West Chester, PA, USA
e-mail: j.fava@pe-international.com

A. Smerek
PE INTERNATIONAL, Inc., Ottawa, ON, Canada

A. B. Heinrich
Heidelberg, Germany

L. Morrison
PE INTERNATIONAL, Inc., Boston, MA, USA

W. Klöpffer (ed.), *Background and Future Prospects in Life Cycle Assessment,*
LCA Compendium – The Complete World of Life Cycle Assessment,
DOI 10.1007/978-94-017-8697-3_2, © Springer Science+Business Media Dordrecht 2014

The general mission of the SETAC LCA Groups is to proactively advance the science and application of LCAs to reduce the resource consumption and environmental burdens associated with products, packaging, processes or activities.

Although life cycle assessment promised to be a valuable tool in evaluating the environmental consequences of a product, process, or activity, the concept was relatively new and required a framework for further development.

The workshop, 'A Technical Framework for Life Cycle Assessments', held August 18–23, 1990, at Smugglers Notch, Vermont, was organized by SETAC to develop a framework and consensus on the current state of LCA and research needs for conducting life cycle assessments. Although life cycle assessments have been used, in one form or another, before the name was coined, this workshop report is the first document which presented the name of the method.

The four SETAC LCA workshops in Smugglers Notch (1990), Leiden (1991), Sandestin (1992) and Wintergreen (1992) formed a tiered process to culminate in the Code of Practice workshop of Sesimbra, Portugal, March 31–April 3, 1993.

Developing international consensus on harmonized methods has been a goal of the SETAC LCA workshops. The 'Code of Practice' completed the harmonization process. Shortly after the workshop, during the autumn of 1993, the ISO standardization process was initiated.

In 1994, as a result of the SETAC LCA workshops, the LCA Advisory Group of SETAC and the LCA Steering Committee of SETAC Europe established individual work groups to address specific LCA issues.

SETAC's working groups and workshops have advanced both the application and reputation of Life Cycle Assessment (LCA) by authoring LCA publications, supporting the development of LCA standardization, partnering with United Nations Environmental Programme (UNEP), and advancing the use of LCA in various sectors. As SETAC grows and expands on its own and with its supporters and partners, it will continue to advance the understanding and use of LCA while ensuring that science is kept at the forefront of LCA development.

Keywords Global coordinating group (GCG) • International organization for standardization (ISO) • LCA in developing countries • LCA in the building sector • Life cycle assessment (LCA) • Pellston workshops • SETAC Europe LCA steering committee • SETAC LCA groups • SETAC North American LCA advisory group • UNEP/SETAC Life cycle initiative • Work groups life cycle impact assessment •Work groups simplified/Streamlined LCA • Workshop Leiden • Workshop Sandestin • Workshop Sesimbra • Workshop Smugglers Notch • Workshop Wintergreen

1 Introduction—SETAC and Life Cycle Assessment

Google labs' Books Ngram Viewer allows any user to graph the frequency of occurrence of words or phrases in Google's database of 500 billion words from digitized books. That technology enticed the senior author to investigate the relationships of a number of Life Cycle Assessment (LCA)-related words over time. One combination, 'SETAC' (Society of Environmental Toxicology and Chemistry) and 'life cycle assessment,' yielded a very interesting observation for the years 1980–2008.

The 'SETAC' acronym first appeared in books in the 1980s. The frequency of appearance grew steadily from 1990 through 2004, showing a tenfold increase. In 1990, SETAC sponsored an international workshop where the term 'life cycle assessment' was coined. SETAC subsequently established the accepted name (and framework) for life cycle assessment (Fava et al. 1991). Previously, a few practitioners in the United States and Europe used different terms such as 'Resource and Environmental Profile Analysis' (REPA) (Hunt and Franklin 1996).

The occurrence of the phrase 'life cycle assessment' in books grew very similarly to the occurrence of 'SETAC' from 1990 through 2004. Was this a coincidence or were there activities within SETAC that contributed to this parallel growth? As growing interest in green buildings and sustainable products (to name a few drivers) increased the use of LCA, a review of the recent history behind SETAC's role was required. Klöpffer (2006) provides an excellent summary of the role of SETAC in the development of LCA; and Ekvall (2005) outlines the further advancement of LCA by SETAC's LCA working groups.

2 Life Before SETAC's Involvement with LCA

2.1 Focus on Pollution Reduction

In 1969, the Cuyahoga River in the United States became infamous for being 'the river that caught fire'. This event helped spur the environmental movement. The river burned because of pollution dumped in it by nearby industrial and waste water operations. At the time, there were few environmental laws providing direction or restriction of environmental releases for companies. The river that caught fire became a national symbol of the fundamental flaws in the way society treated the environment.

Laws and regulations were instituted in the early 1970s that placed new and/or additional controls on point-source releases of waste from treatment facilities and industrial operations. As a result, the water quality of the rivers improved. There is still much to learn about the risks of ingredients and emissions from our products and processes that enter our rivers and waterways, but significant progress has been made.

These governmental and regulatory expectations, placed on companies and government behavior, primarily related to the management of emissions and waste from manufacturing operations (and later the cleanup of abandoned or contaminated land). They were instrumental in creating a change for improved environmental management.

In the 1980s, many regulatory approaches to environmental protection continued to be based on 'end-of-pipe' solutions that focused on a single medium (e.g., air, water, or soil), a single stage in the product's life cycle (e.g., production, use, or disposal), or a single issue (e.g., individual chemical limits). Such strategies did not always lead to a net environmental benefit. Environmental laws and regulations that have a single focus often force the use of pollution control resources in ways that are not optimal for reducing overall impacts.

The attempt was made to solve a single environmental problem without considering the interconnectivity of natural systems. Designed legislation, although intended for a specific purpose, has regularly created additional, unexpected environmental problems. Single-issue approaches are often not designed with a systematic understanding of the tradeoffs and their implications. Thus, they frequently diminish opportunities for achieving net environmental improvements.

One of the rapidly evolving landscapes in business today is adaptability to the changing nature of environmental impact management. This occurs as scope expands from a single site and/or issue to a full understanding of the impacts of our products over their entire life cycles. Many advertisements pitch 'green' product traits, but all products have environmental impacts. Materials and crude oil are extracted from the earth, processed, combined with other materials to make parts, assembled into finished products, shipped to customers, and ultimately delivered to final consumers who use the products and dispose of them. Along that value chain, energy is used, waste is generated, and more natural resources are consumed. Sustainability will require us to continue creating value for society while reducing environmental and social impacts.

2.2 *Moving Beyond Pollution Control to Pollution Prevention*

With the improvement in the treatment of air and water emissions and waste from manufacturing operations, there was recognition that end of pipe treatment can only go so far. Additional examination of what enters the end-of-pipe treatment was needed. This led to the development of pollution prevention and pre-treatment programs. These programs were not as influenced by explicit government regulations. It did become clear, at least in some groups, that preventing pollution from entering the environment could save the organization money and protect the environment. 3M's Pollution Prevention Pays (3P)program, was a landmark program initiated in the 1980s which has saved 3M 1.2 billion dollars worldwide. As well, it has prevented 2.9 billion pounds of pollutants from entering the environment.

In the 1970s and 1980s, there were a number of studies and situations that created the demand for additional information on environmental impacts of products. These were primarily driven by solid waste management issues. Three in particular are relevant to this conversation[1]: (1) duelling diaper debates; (2) mercury in fluorescent light bulbs; and (3) Coco-Cola demanding supply chain improvements.

[1] These were presented in Fava (2012) Life cycle knowledge informs greener products, Chap. 25, in: Curran MA (ed) LCA Handbook—a guide for environmentally sustainable products.

2.2.1 Duelling Diaper Debates

Many of us remember the garbage barge that went up and down the east coast of the United States in the late 1980s looking for a disposal site. In this period, there were concerns about the significant amount of solid waste that society was generating. Today that concern remains, but society has realized that there is a broad and growing array of environmental issues. A study by the cloth diaper industry revealed that the use of cloth diapers did not create as much solid waste as the use of disposable diapers (now called single-use diapers). Subsequently, there was a push to use more cloth diapers and reduce the number of single-use diapers sent to landfills. However, additional studies, using methods including life cycle assessment, showed that cloth diapers also have meaningful environmental impacts during use (e.g., heating water for washing). It became unclear which product was actually better. The 'Dueling Diaper' LCA studies raised awareness of the diversity of environmental impacts that products can create and the environmental trade-offs between product options.

What did we learn? One of the most significant lessons was the realization that, depending upon the impact in question and where it occurs, different and equally valid interpretations can result. These early studies revealed that all products have impacts on the environment. LCA tools enable decision makers to use new and additional information on multiple metrics to make better-informed decisions. A clear recognition of the importance of continuing to ensure performance of the product is maintained and improved.

2.2.2 Mercury in Fluorescent Light Bulbs

Society and policy makers were faced with demand to reduce mercury levels associated with lighting systems in order to reduce the overall release of mercury into the environment. While incandescent bulbs contain no mercury, mercury is a critical element in fluorescent bulbs that increases efficiency and durability. The resulting reduction in energy consumption causes a corresponding drop in mercury emissions from coal-fired power plants. Due to the concern about mercury entering the environment from landfills, policy makers were wrestling with two options: banning fluorescent lamps from municipal solid waste (MSW) facilities or encouraging greater use of fluorescent lamps over incandescent lamps. If we only consider the amount of mercury that might enter the environment as a result of bulb disposal, it is clear that significantly more mercury would come from florescent bulbs because incandescent bulbs don't contain any mercury.

However, if we expand the system boundaries to include the use phase of the bulbs in addition to their disposal, what does the data reveal? Surprisingly, we find that the use and disposal of incandescent lights released into the environment, on average, *four to ten times as much mercury* as the use and disposal of fluorescent lights. This is due to the additional power plant emissions created by the inefficient incandescent bulbs during the use phase. The US Environmental Protection Agency

(EPA) estimated, in the early 1990s, that the use of fluorescent lights will also eliminate the following:

- 50 % of aggregate national electricity demand;
- 232 Million t of CO_2 emissions each year;
- 1.7 Million t of SO_2 emissions each year; and
- 0.9 Million t of NO_2 emissions each year.

Clearly, when the system boundaries and the impacts of interest are expanded to include bulb use as well as disposal, the better decision is to encourage greater use of fluorescent bulbs. This was, in fact, the direction taken by policy makers: *use fluorescent bulbs* but challenge lighting companies to *reduce the mercury in those bulbs.* This policy enabled reduction of mercury releases and encouraged innovation to develop lighting systems containing less mercury. Philips' Sustainable Lighting Solutions and its ALTO® bulbs[2] are good examples of environmentally responsible lighting because they contain less mercury, are Toxicity Characteristic Leaching Procedure (TCLP[3]) compliant, energy efficient and last longer. Responsible lighting solutions result in fewer light bulbs in landfills and further reduce the impact on the environment.

What did we learn? By taking a broader systems approach, we can make better decisions. Using this systems thinking, the lighting sector has produced innovations over the years to develop better products with lower environmental impacts.

2.2.3 Coca-Cola's Supply Chain Improvements

In the early 1970s, the Coca-Cola Company conducted a study of its beverage containers. The results showed that all of their beverage containers had real environmental impact. In response, Coca-Cola decided to challenge the material and container companies to make adjustments to their products and processes that would result in reduced life cycle environmental impacts over previous design options. This was contrary to common practice at the time to simply ban or deselect the poorest-performing material(s)[4]. For Coca-Cola's aluminum cans, the sector worked with local governments to develop a recycling infrastructure for the used beverage containers, resulting in a reduction of more than 90 % in the energy used throughout the life cycle of the aluminum beverage container. Other material groups made similar improvements in a response to Coco-Cola's challenge.

What did we learn? LCA study results should be used to improve product environmental performance. As Coca-Cola chose not to ban any of the high environmental

---

[2] http://www.usa.philips.com/c/fluorescent-tubes/296298/cat/en/.

[3] The Toxicity Characteristic Leaching Procedure (TCLP) is designed to determine the mobility of both organic and inorganic analytes present in liquid, solid, and multiphasic wastes. http://www.epa.gov/osw/hazard/testmethods/sw846/pdfs/1311.pdf.

[4] Information based on personal conversations with Coca-Cola employees.

burden materials, they created an atmosphere that allowed for material innovation such as the development and financing of a recycling infrastructure to recapture the inherent value in the aluminum.

3 The Birth of SETAC

Rachel Carson's 'Silent Spring'[5] was published in 1962 and the world was awakened to the implications of wide spread chemical and pesticide use on the environment. Federal laws around the world were passed that began to demand additional information on the fate and effects of chemical use on the environment. This information was to inform the determination related to what might be acceptable and safe use and application rates for chemicals and pesticides used in commerce. However, this created a need for internationally acceptable methodologies and risk assessment frameworks. These frameworks could be applied and used to examine and evaluate the safe and acceptable levels of chemical use. Although there was a demand for environmental health data on chemicals, there was no global scientific organization able to talk about the science behind the regulations being developed.

In 1979, The Society of Environmental Toxicology and Chemistry (SETAC) was founded to serve as a non-profit professional society to promote the use of multi-disciplinary approaches in the study of environmental issues. SETAC has three unique strengths: its global scale, its tripartite[6] membership and governance and its scientific base. Because SETAC was developed on an international scale, it has been able to address global environmental issues. In October 1980, there were 230 Charter Members. Today, there are nearly 6,000 members from more than 100 countries. SETAC's members, from governmental, academic, and business backgrounds, are committed to balancing the scientific interests of the three sectors represented.

3.1 SETAC Workshops

From the beginning, SETAC has sponsored workshops to bring together scientists, engineers, and managers from government, private business, academia, and public interest groups to consider the state-of-the-art of specific environmental topics. While formats vary, workshops are generally held over the course of 4–5 days with 40–50 individuals in attendance. During the intensive workshops, a combination of formal presentations and informal working sessions are used to examine the status of current information and the knowledge base of the topic and develop recom-

[5] Carson (2002) [1st Pub. Houghton Mifflin 1962]. Silent Spring. Mariner Books.*ISBN 0-618-24906-0*. Silent Spring initially appeared serialized in three parts in the June 16, June 23, and June 30, 1962 issues of *The New Yorker* magazine.

[6] SETAC has a commitment to balance the scientific interests of government, academia and business.

mendations for enhancing the current state of the science. An expected product of a SETAC workshop is a document that presents a clear description of this knowledge and a description of the recommendations developed.

There are two general categories of SETAC Workshops: Pellston Workshops, named after the location of the first workshops (University of Michigan Field Station, Pellston, MI, USA) and Technical Workshops, including Small Meetings. The distinction between these two categories is primarily a function of the anticipated breadth of interest in the topic across the SETAC membership, the criticality and timeliness of the topic, and the likelihood that the workshop will significantly advance scientific understanding of the issue. The basics of balance and objectivity underlying all SETAC activities apply to all SETAC workshops. This foundation has been the key to the successful workshops for the past 30 years on topics ranging from LCA to improving management of contaminated sediments.

All SETAC workshops must adhere to the following fundamental guidance principles:

- Proposed workshop is consistent with SETAC goals;
- Proposed content is scientifically sound or credible;
- Workshop promotes multi-disciplinary approaches;
- Workshop attendance ensures balance in opinion and representation by involving relevant constituencies (from academia, government, business and non-governmental organizations);
- Workshop has a viable communication plan coordinated with the SETAC Publications Advisory Council (PAC) and the Public Relations and Communication Committee (PRCC) that ensures timely, accurate and cost-effective publication of results to the Society and sponsors as well as to a wider global audience. Web-based communications, such as webinars, podcasts, or blogs, are also encouraged to the extent they increase the base of knowledge of the workshop findings and complement other, more traditional means;
- Proposed workshop does not generate a conflict of interest; and,
- Workshop budget is viable, including publication costs, and financial liability to SETAC is controlled.

3.1.1 Pellston Workshops

The goal of a Pellston Workshop is to promote advancement in the resolution of truly cutting-edge technical and policy issues in environmental science, while enhancing strategies of science and philosophy.

Developing the Workshop Topic Workshop proponents, at least one of whom must be a SETAC member or an individual SETAC member, will identify a pertinent issue or environmental topic to serve as a focal point for the proposed workshop and form a tentative Workshop Steering Committee which includes at least one SETAC Office staff member as an ad-hoc member. The tentative Steering

Committee or individual will develop a pre-proposal (concept paper) that clearly describes the topic to be examined, the workshop objectives, the anticipated range of participants, any potential funding sources, and the benefits to be obtained from the workshop. The SETAC Office should be contacted for example documents, for advice on preliminary workshop organization, for ideas on tentative dates and location, and for funding requirements for the workshop and follow-up activities. If the Board/Council feel that the pre-proposal for the workshop is worthy, it will approve preparation of a complete proposal. This proposal is submitted to the Technical Committee, through the SETAC Office, for technical review.

Technical Committee Review After referral from the Board of Directors/Council, the appropriate national SETAC Technical Committee would review the proposal in reference to the *Criteria for Designation of SETAC Pellston Workshops* and *Criteria for Designation of SETAC Technical Workshops.* Based upon this review, the Technical Committee would provide a recommendation to the Board of Directors/Council on the degree to which the workshop should be sponsored[7].

Based on the recommendation of the relevant Technical Committee and in light of other demands on the SETAC Office and the probability of obtaining adequate funding for the proposed workshop, the Board, Council, or SETAC World Council (SWC)[8] would approve or disapprove the workshop/meeting or the SWC would refer back to national level as a technical workshop as relevant. This would take place at any regularly scheduled meeting throughout the year or by ballot.

Planning the Workshop After approval by the SETAC Board of Directors, workshop proponents may initiate the planning process, cooperatively with the SETAC/SETAC Foundation Office.

Types of Publications From Workshop Proceedings Publication and dissemination of proceedings from workshops is highly encouraged. The outline and format for the publication is dependent upon the workshop objectives and program. Development of a complete first draft of the proceedings during the workshop is essential. After the workshop is held, it is the responsibility of the Workshop Steering Committee to ensure that the proceedings are completed.

Success of Pellston Workshop While SETAC was founded to promote the use of multi-disciplinary approaches for the study of environmental issues, the format of the Pellston Workshop laid the foundation necessary to address life cycle assessment (LCA).

[7] The following recommendations are possible: Recommend SETAC sponsorship as a Pellston Workshop and submission to the SWC Technical Committee for consideration; Recommend SETAC sponsorship as a Technical Workshop at a national level; Recommend SETAC sponsorship as a Small Meeting at national level; Recommend SETAC sponsorship contingent upon securing funding; Recommend SETAC sponsorship contingent upon incorporation of mandatory changes; Recommend SETAC sponsorship with minor changes suggested; or, Recommend against SETAC sponsorship.

[8] The SWC facilitates worldwide outreach to environmental scientists, engineers, and managers and encourages development of additional SETAC member groups.

3.1.2 Technical Workshops

SETAC supports the convening of technical workshops to bring together experts to discuss and resolve timely technical, scientific or policy issues related to environmental science. SETAC's level of support can range from simply providing an endorsement (e.g., non-exclusive license to use SETAC name or logo for promotional purposes) to providing full technical and scientific support, as long as basic principles are met. Recognizing the diversity of possible workshop formats and varying levels of potential logistic and financial involvement, SETAC's approval and sponsorship criteria are generally flexible and determined primarily by the level of support sought by the applicant. The governing principle: the greater support that is requested from SETAC (in terms of support and financial commitment), the more detail and oversight will be required from the applicant.

Among the flexible workshop formats, there are three general levels of SETAC involvement:

- *Level 1—SETAC-hosted technical workshop or meeting:* Major workshop, hosted by SETAC, of global, regional or national relevance on an important, but not necessarily urgent environmental scientific or policy issue (i.e. non-Pellston), organized and advertised by a SETAC-assembled Steering Committee, with all-invited attendance, significant scientific input, major SETAC financial and logistical support and a substantive high-quality publication.
- *Level 2—SETAC-co-organized technical workshop:* Workshop of global, regional or national relevance, co-organized by SETAC (in partnership with other organizers), with significant scientific input, some SETAC financial and logistical support, but limited financial liability and less comprehensive publication effort.
- *Level 3—SETAC sponsored technical workshop or meeting:* Workshop is organized by a different organization, but endorsed, co-sponsored or advertised by SETAC, with a certain degree of scientific input, but minor or no financial and logistical support.

The goal of a SETAC-hosted or co-organized technical workshop (Level 1 or 2) is to promote advancement of the resolution of important technical and policy issues in environmental science while enhancing strategies of science and philosophy.

To that end, the following criteria have been developed as guidance for the designation of SETAC Technical workshops (in addition to the general guidance principles listed above):

- The proposed workshop topic does not merit a Pellston workshop designation;
- Potential sources and estimated amounts of funding are clearly identified;
- SETAC member participation in balanced Steering/Organizing Committee is required;
- Steering Committee Chair or Co-Chair is a strong champion for the proposed workshop. Members of the steering committee must include recognized subject matter experts germane to the workshop topic;

- Steering Committee and workshop participants must represent an internationally or nationally diverse group of individuals representing academia, government, business, and other non-government organizations;
- Workshop topic is recognized as being an important, although not necessarily pressing, scientific issue by the Science/Technical Committee and Council/Board of the host SETAC unit, and is expected to be of significant interest to a reasonable number of people in the environmental community;
- Workshop objectives are clearly identified and the workshop is designed so that important aspects of the topic can be addressed and definitive conclusions/action items can be developed within the time frame of the workshop;
- Proposed workshop products have a high likelihood of contributing to our understanding of an important issue in environmental science and will be recognized as being of value to specific sectors within the scientific community (i.e., establishing what is known, where uncertainties exist, what research is needed to address those uncertainties);
- An adequate publications and communications plan has been formulated that includes at a minimum an Executive Summary document, a SETAC Globe article and a Presentation at an annual SETAC meeting. In addition, weblog report, webinar or podcast is strongly recommended to further disseminate outcomes; and,
- The potential for conflict of interest does not exist or is acceptably resolved.

4 Early Days of SETAC 1990–1993

4.1 SETAC LCA Groups

The SETAC North American LCA Advisory Group is a formally recognized group within SETAC that has been in existence since June 1991. Similarly, SETAC Europe established an LCA Steering Committee. Both the LCA Advisory and Steering Committee are referred to as the SETAC LCA Groups. The following provides an overview of the initial organization and roles of the SETAC LCA Groups[9].

The LCA Groups report to the Board of Directors of both SETAC and SETAC Europe. Specific activities such as workshops, conferences, or educational material development, including 'position papers', are approved by the Board of Directors. During the 1990s these SETAC LCA Groups were instrumental in driving the scientific progress to codify the professional practice of LCA. During this time period, several major workshops were successfully organized and over a dozen key publications produced. The SETAC LCA Groups also broadly supported the initial

[9] Although the specific details between the two SETAC LCA groups may be slightly different, the intent of this section is to describe the role and value of the SETAC LCA groups in advancing LCA within SETAC generally.

preparation of the ISO 14040 series of voluntary international standards as well as their subsequent revisions (see Sect. 5 for more information).

SETAC LCA Group Organization and Structure SETAC guidelines call for an LCA Group to organize itself around a basic structure consisting of a leadership group, rank and file members. The leadership consists of elected LCA Groups between 6 and 15 members, each. The LCA Group composition should reflect the tripartite (business, academia, government) balance that SETAC tries to achieve in its operating and membership components. For each LCA Group, a chair is identified from among the elected LCA Group members and recommended for appointment to the SETAC North America and/or SETAC Europe President. Once appointed, this individual will serve a three year term. Other officers (co-chair, communications officer, etc.) of each LCA Group may be designated by the LCA Group as appropriate. Historically, the key to an active and effective LCA Group leadership has been engaging the entire LCA Group members rather than placing the majority of the burden on the chair to organize and run the group.

When initially organized, all LCA Groups prepare a Standard Operating Procedure (SOP) that describes their mission, functional organization, and objectives. In addition, the SOP calls out topics or issues within the purview of the Group and activities, including various communications activities, in which they will be engaged to achieve their goals.

When an LCA Group identifies a number of technical activities within their overall topic, ad-hoc work groups are typically set up to address those issues. Though not a requirement, often the LCA Group members will lead those working groups. For other functions not formally assigned to the LCA Group or taken up by an ad-hoc working group and which are recurring, a standing committee may be formed. Types of activities for which this may be appropriate include, but are not limited to: annual meeting session planning and execution, short courses, webinars, fundraising, etc.

Mission of LCA Groups The general mission of the SETAC LCA Groups is to proactively advance the science and application of LCAs to reduce the resource consumption and environmental burdens associated with products, packaging, processes or activities. To achieve this mission, the LCA Group[10] will:

- Serve as a focal point to provide a broad-based forum for the identification, resolution, and communication of issues regarding LCAs; and,
- Facilitate, coordinate, and provide guidance for the development and implementation of LCAs.

[10] Each LCA Group may have a slightly different mission but generally the purpose is to advance the science and development and application of LCA.

The LCA Group's success in meeting its mission depends on the willingness of its members to voluntarily identify, initiate and conduct activities. At this time, several interest topics have been identified by the current Advisory Group[11].

As noted above, the SETAC LCA Groups are a recognized entity within the SETAC organizational structure. When the proposition was put forth to recognize LCA Groups, it was with the anticipation they would become forums for advancing activities within the Society around certain professional interest areas. In addition, within a 'bottom-up' organization such as SETAC, a geographically oriented Group could deal with regional issues and at the same time represent those regional perspectives on a global stage. It was expected that, given the concentration of technical expertise, the SETAC LCA Group would speak out officially in the name of the Society on occasion within the topical coverage of the Group. In addition to serving its members, a primary goal of SETAC is to provide balanced, scientific information to planners, legislators, managers, regulators, and others. It would further assist in the development of technically sound environmental policies, laws, and regulations.

4.2 LCA Group Activities

SETAC's LCA Groups have successfully held workshops and conferences and have developed pertinent educational material, including 'position papers'. The following sections outline several of the LCA groups' workshops and successes.

4.2.1 A Technical Framework for Life Cycle assessment. August 18–23, 1990, Smugglers Notch, Vermont

Although life cycle assessment promised to be a valuable tool in evaluating the environmental consequences of a product, process, or activity, the concept was relatively new and required a framework for further development.

The workshop, 'A Technical Framework for Life Cycle Assessments' (Fava et al. 1991)', held August 18–23, 1990, at Smugglers Notch, Vermont, was organized by SETAC to develop a framework and consensus on the current state of LCA and research needs for conducting life cycle assessments. Although life cycle assessments have been used, in one form or another, before the name was coined, this workshop report is the first document which presented the name of the method.

The workshop involved 54 scientists and engineers of diverse technical backgrounds representing governmental organizations, universities, industries, public

[11] Recent topics of interest that are being considered for the Advisory Group, include: US Green Building Council's LEED program to identify sustainable buildings; US Database project to make inventory data publicly available; Creation of LCA sessions at the annual SETAC meeting; Liaison with the various task forces within the Life Cycle Initiative; Development of an awards program to recognize exemplary contributions in the field; and, to Identify opportunities for capacity building in developing countries.

interest groups, consultants, and contract research firms. Also, participants were invited from Europe, Japan, and Canada.

The workshop focused on defining concepts and developing a framework for the inventory component of an LCA. However, it also identified the need to conduct particular workshops to evaluate other LCA components.

Workshop Objectives 'A Technical Framework for Life Cycle Assessments' workshop objectives were:

- to clarify definitions and terms associated with life cycle assessments;
- to provide a forum for information exchange among researchers from government, industry, academia, and public interest groups; and,
- to agree on a technical framework of key life cycle assessment components.

The charge given to workshop participants was agreement on a technical framework of key life cycle assessment components and identification of the research needed to improve life cycle assessment techniques.

Workshop Format During the initial phase of the workshop, keynote presentations on the development and use of life cycle assessments were made by individuals representing SETAC, U.S. Environmental Protection Agency, Environmental Defense Fund, state governments, and industries. The objective of this initial phase was to establish a common information base.

Prior to the workshop, each participant was asked to prepare a list of issues and thoughts related to improvement of understanding and development of life cycle assessments.

Participants were placed in one of six workgroups: Raw Materials Acquisition; Processing, Manufacturing, and Formulation; Distribution and Transportation; Use/Re-Use/Maintenance; Waste Management; and Integration.

Report Organization The workshop report presents a general technical framework from which specific methods and procedures could be developed.

Chapter 1 presents an overview of the technical framework for life cycle assessments and a historical perspective on life cycle assessments. Chapter 2 provides an overview of the framework for life cycle inventories. Specific discussions on aspects of the life cycle inventory component (Component I) are presented for Raw Materials and Energy (Chap. 3); Manufacturing, Processing, and Formulation (Chap. 4); Distribution and Transportation (Chap. 5); Use/Re-Use/Maintenance (Chap. 6); Recycling (Chap. 7); and Waste Management (Chap. 8). The research needed to improve the inventory component of a life cycle assessment is discussed in Chapter 9. Research directions and technical considerations necessary to advance the technical framework into Component 2 (Impact Analysis) and Component 3 (Improvement Analysis) are discussed in Chapter 10.

Appendix A is a glossary of the technical terms used in the workshop report. Appendix B is a complete list of participants in the workshop, and Appendix C contains references and a bibliography of reports on Life Cycle Assessment.

Main Findings One of the major findings of the workshop was consensus that complete life cycle assessments should be composed of the following three separate but interrelated components:

- Life cycle inventory
- Life cycle impact analysis
- Life cycle improvement analysis

The three components represented the first attempt to develop a structure for a life cycle assessment and provided the information needed to maximize environmental improvements. This structure has been traded as 'SETAC triangle'.

The existing life cycle assessments focused on the inventory component. As such, most of the participants addressed the life cycle inventory component of a life cycle assessment. Therefore, considerable research was necessary to develop the impact and improvement analysis components.

Participants developed a technical framework for the key phases of a life cycle inventory. The major life cycle inventory stages are (1) raw materials acquisition; (2) manufacturing, processing, and formulation; (3) distribution and transportation; (4) use/re-use/maintenance; (5) recycling; and (6) waste management.

One major finding was related to the question of aggregation of individual environmental release quantities. The workshop participants agreed that the summation of dissimilar materials in the life cycle inventory is scientifically unjustified and represents an incorrect technical approach in the inventory component of a life cycle assessment. However, it was agreed that some summations are possible; for example, summing the same pollutant emitted from different sources, but in the same form and to the same sector of the environment. Also, some categories of data (i.e., solid waste, energy consumption) can be summed (as long as individual data are also provided).

While confidential or proprietary information must be protected, the workshop concluded that methods and data from a life cycle inventory should be available for public review if the document is to be used in the public domain in a decision-making context.

Presentation of quantitative data should include an identification of data sources and the extent of data completeness and variability. Whereas, categorization of data may be employed, aggregation of data should be avoided whenever feasible, and dissimilar data should not be aggregated. It was recommended that a review of national and international standards and other possible conventions be undertaken to generate general guidelines for data grouping.

Research Needs The following specific research needs to improve the life cycle inventory methods were identified during the workshop:

- *Database development,* including: the development of data quality standards; development of generic databases and guidance on when and how they should be used; evaluation of how industry average data should be used in life cycle inventories; and, development of additional databases.

- *Inventory methodology refinement,* including: Criteria and applications guidance to determine what level of input and output data is meaningful; establishment of a standard list of waste sources and pollutants; development of generic models; development of approaches to allocate inputs and outputs among co-products; development of approaches to allocate energy and environmental releases among incoming waste streams and to all environmental media; development of approaches to incorporate data variability; development of approaches to take into account sensitivity analysis in life cycle inventory methodology; establishment of a peer review process; standardization of life cycle inventory methods; and, development of effective approaches for communicating life cycle inventory results.

Recommendations The following recommendations came from 'A Technical Framework for Life Cycle Assessments' workshop:

- A multiyear research initiative is needed to ensure the continued development of effective life cycle assessment strategies and methods;
- Initial efforts should focus on refining thelife cycle inventory component;
- Additional efforts should include development of approaches to help progress beyond the inventory to the impact and improvement analysis components of a life cycle assessment;
- Sufficient case studies should be developed to demonstrate the usefulness of life cycle assessment methodology to a wide range of products, processes, and activities;
- The research initiative should be expanded to include applications of the life cycle assessment methods to illustrate their use in actually improving products, processes, and activities; and,
- This new research initiative should build upon and enhance relevant existing pollution prevention research activities.

4.2.2 Life Cycle Assessment: Inventory, Classification, Valuation, and Data Bases. December 2–3, 1991, Leiden, The Netherlands

One month after the workshop at Smugglers Notch, Vermont (August 1990), a European workshop took place in Leuven, Belgium, September 24–25, 1990, on 'Life Cycle Analysis for Packaging Environmental Assessment' (SETAC Europe 1990). Due to the increasing problems of waste disposal, packaging was the main topic of the existing LCAs. The workshop followed a similar aim as the workshop in Smugglers Notch, which was to bring together the groups working on life cycle based assessment methods. This was necessary on both sides of the Atlantic Ocean, since the methods were not really new, but uncoordinated and distanced from harmonization as well as standardization (Klöpffer 2006).

The Workshop in Leiden, The Netherlands The workshop in Leiden, The Netherlands, was chaired by Helias A. Udo de Haes, Centre of Environmental Science

(CML), Leiden University. Approximately 50 participants from Europe and North America attended.

During the Leiden workshop, a discussion occurred about the general set-up of environmental LCA.

First, it was concluded that the term LCA can best be interpreted as 'life cycle assessment' instead of 'life cycle analysis', including both objective and normative steps. Second, different sections of the environmental LCA procedure were identified: goal definition, inventory, impact analysis, valuation. They contribute to the 'improvement options'.

Significantly, the following issue was raised at the workshop: the class of studies called life cycle assessment of products (LCA) might be subdivided into the following interrelated subclasses:

- environmental life cycle assessment
- economic life cycle assessment
- social life cycle assessment

Today, sustainability is discussed as the ultimate goal of LCA. In the 'three pillar' interpretation of sustainability, environmental, economic and social aspects have to be considered. In life cycle product assessment, LCA deals with the environmental aspects only. For the complete assessment, however, the economic and social life cycle aspects have to be included as well (Finkbeiner et al. 2010; Klöpffer 2003, 2008; O'Brien et al. 1996; Valdivia et al. 2012). The 'three pillar' concept of sustainability is often called the 'triple bottom line'.

$$\text{LCSA}=\text{LCA}+\text{LCC}+\text{SLCA}$$

LCSA: Life Cycle Sustainability Assessment
LCA: (environmental) Life Cycle Assessment
LCC: (environmental) Life Cycle Costing
SLCA: Social Life Cycle Assessment

Results. The workshop showed that the LCA methodology was acknowledged to some extent, but confidence and experience with it were still lacking.

4.2.3 A Conceptual Framework for Life Cycle Impact Assessment. February 1–7, 1992, Sandestin, Florida

To develop a consensus on the state of the practice and research needs for conducting life cycle impact assessments, approximately 50 experts in LCA and environmental impact assessment assembled for a 1-week workshop. The workshop was held February 1–7, 1992, in Sandestin, Florida, USA.

The Life Cycle Impact Assessment Workshop marked the first time that the SETAC offices in Europe and the United States shared responsibility in identifying and bringing together international experts for a Pellston workshop (the workshop was the twelfth in a series of Pellston workshops). The participants represented state and federal agencies, industry, universities, public interest groups, and research laboratories in the United States, Canada, United Kingdom, Belgium, Denmark, France, Germany, and The Netherlands.

The Workshop The workshop objectives were to define impact assessment in the context of an LCA, to discuss and develop a consensus by what means impact assessments could be applied to LCAs, and to assess the overall need for developing feasible impact assessment methods for LCAs. In addition, research needs were identified to improve the impact assessment component of LCAs.

The workshop followed a three-phase format. During the initial phase, discussion initiation papers were presented covering three general areas: LCA background, life cycle impact assessment approaches, and impact assessment methodology. During phase 2, small work group sessions identified and discussed impact categories. During phase 3, the individual work groups[12] identified and discussed the existing impact assessment methods, their potential applications to LCAs, and research needs.

Prior to the workshop, each participant was asked to prepare a list of issues and thoughts related to improving our understanding and developing the life cycle impact assessment component.

Each work group was responsible for developing a summary of the findings discussed during the week. The Steering Committee was responsible for synthesizing the findings of the work groups into a unified report.

Objectives The workshop participants were charged with defining impact assessment in the context of life cycle assessment. Additionally, they were asked to discuss and develop a consensus on whether and by what means existing impact assessment tools could be applied to LCAs. For those areas where consensus could not be reached, the participants were asked to identify research needs to improve the impact assessment component. Although the impact assessment component is still in an early stage of development, a number of existing impact assessment tools were identified that might be applied to LCAs.

Major Findings It was agreed that conducting an LCA is not a linear process but one that incorporates feedback loops and requires interaction among the LCA components. The workshop participants reaffirmed the value of the three-component model for LCAs developed at the Smugglers Notch workshop in 1990. Also, building on the results of the Leiden workshop in 1991, a goal definition and scoping component was incorporated as an additional step which would serve to specify the purpose and expected products of the study, select boundaries, define assumptions, and determine what is to be included or excluded consistent with the goal of the study. Figure 2.1 shows the amended 'SETAC triangle' 1992. Differing from the workshop in Leiden in 1991, the improvement component was included in the technical framework.

[12] The six work groups were human health, ecological (chemical) stressor, ecological (nonchemical) stressor, resource depletion, valuation, and integration.

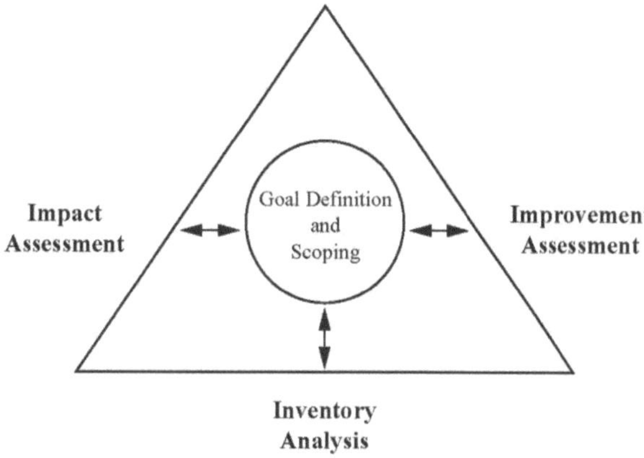

Fig. 2.1 SETAC triangle 1992 (Sandestin workshop) and 1993 (Sesimbra workshop) until the ISO standardization process when 'Improvement assessment' was replaced by 'Interpretation'

The major impact categories were defined. The primary impact categories were human health, ecological health, and resource depletion[13]. Other impact categories were human health, ecological health, and resource depletion impacts associated with changes in social welfare aspects. Occupational health considerations were included within the human health category.

Based on discussions at the workshop and during the SETAC Europe workshop in Leiden, the workshop participants agreed to a three-step conceptual framework for impact assessment as follows:

1. *Classification*—The process of assignment and initial aggregation of data from inventory studies to relatively homogenous stressor categories (e.g., greenhouse gases or ozone depletion compounds) within the larger impact categories (i.e., human and ecological health, and resource depletion).
2. *Characterization*—The analysis and estimation of the magnitude of impacts on ecological health, human health, or resource depletion for each of the stressor categories, derived through application of specific impact assessment tools.
3. *Valuation*—The assignment of relative values or weights to different impacts and their integration across impact categories to allow decision makers to assimilate and consider the full range of relevant impacts across impact categories.

This three-step impact assessment model further developed the two-step impact assessment model discussed during the Leiden Workshop. One of the significant findings was the importance of the stressor concept to bridge the gap between the inventory and impact assessment components. A stressor was defined as a set of conditions that may lead to impacts. The valuation phase, which assigns value or relative weights to the various impact categories, was judged to be inherently subjective.

[13] In today's terminology, primary impact categories are considered Areas of Protection (AoP).

Summary and Future Work The participants recognized that impact assessment is still in an early stage of development and identified a number of research initiatives to enhance the science, practice, and application of LCAs:

- A multi-year research initiative is needed to ensure the development of effective life cycle impact assessment tools and LCA methods in general.
- Case studies should be developed demonstrating the usefulness of the impact assessment steps (i.e., classification, characterization, and valuation), either individually or combined, when applied to a wide range of products, packages, processes, and activities.
- Scoping processes used in other applications should be critically evaluated for their application to LCAs.
- Research is needed to evaluate the cause-and-effect relationship between pairs of stressor-impact linkages relevant to ecological and human health impacts.
- Evaluation of methods to quantify the resource depletion impact category is needed.
- Approaches to applying various decision theory methods to LCAs should be examined.
- The role of social activities and their influence on ecological, human health, and resource depletion impacts should be further considered and approaches to incorporating these impacts in LCAs should be evaluated.

4.2.4 Data Quality: A Conceptual Framework. October 4–9, 1992, in Wintergreen, Virginia

This workshop was the 14th in a series of Pellston-type workshops and the fourth in a series to develop the science, practice, and application of LCAs, in continuation of the workshops in Smugglers Notch (1990), Leiden (1991) and Sandestin (February 1992) (Fava et al. 1994).

To develop a consensus on the state of the practice and research needs for conducting life cycle data quality, approximately 50 experts in LCA and environmental data quality assembled for a one-week workshop. The workshop was held October 4–9, 1992, in Wintergreen, Virginia, USA. The participants represented state and federal agencies, industry, universities, consultants, public interest groups, and research laboratories in Canada, the United States, the United Kingdom, Belgium, Denmark, France, Germany, Japan, Norway, Sweden, Finland, and the Netherlands.

Workshop Objectives The workshop objectives were to:

- identify and understand existing approaches to address data quality issues in LCA studies;
- identify and understand existing approaches to address data quality issues relative to environmental, human health, energy, and resource issues;
- develop a suggested data quality framework considering a distinction among an ultimate framework and steps that can be taken to improve data quality in the

near term and the current state-of-the-practice, recognizing different applications and communication needs;
- develop research needs to improve the quality of LCA data and reduction techniques; and
- begin a process to solicit interest in developing a commitment by users and practitioners to develop LCA integrated international data bases.

Workshop Format As in the case of Smugglers Notch and Sandestin, the workshop followed a three-phase format. The initial phase included the discussion of an initiation paper presented that covered three general areas, namely LCA background, data quality approaches, and data quality assessment methodology. Phase 2 included small workgroup[14] sessions which identified and discussed data quality issues. Phase 3 saw individual workgroups which identified and discussed what data quality assessment methods existed, their potential applications to LCAs, and research needs.

Major Findings Data quality was defined as the degree of confidence with individual input data and in the data set as a whole and ultimately in decisions made by using the data. The reliability of LCA conclusions as final results depends on the quality of the input data and the way they are processed into results using an LCA methodology. The emphasis in this report is on input data.

The workshop provided a strong statement that data quality assessment (DQA) is an integral part of LCA. DQA is a systematic approach to identifying and applying measurements of the suitability of LCA data to meet the intended purpose. Data quality assessment techniques applicable to LCA include the data quality goals (DQGs) process. This process specifies provision for clarity and simplicity in stating data and process related requirements and the establishment of measures of performance to assess data quality.

Use of the DQG process enhances communication, provides a structure for augmenting existing data sets, leads to a focused set of data requirements, and defines the resulting uncertainty of the study results.

The level of data quality achieved is dependent on the level of effort that is allocated to the study, to each subsystem, and to each variable. The appropriate level of effort is influenced by the study purpose, budget and time constraints, data availability, and the need to maintain scientific integrity.

The data quality assessment framework is needed because data in an LCA are diverse and should be assessed by using a logical, formalized, and repeatable method. The framework allows for data documentation in a flexible manner, using quantitative or qualitative measures.

There are many approaches to evaluate data quality. Two were identified as illustrating key issues: qualitative evaluation using a matrix approach, and quantitative evaluation using value trees.

[14] The six workgroups were data quality framework, materials, energy, environmental emissions, ecological health and exposure, and human health and exposure.

In conducting a data quality assessment, it was recommended that sensitivity analysis be used to direct and select the expenditure of time and money to those areas most likely to improve the overall quality of the study. In general, peer review has been used more to ensure the quality of data collection and manipulation procedures than to ensure the quality of the raw data. It was recommended that the current practice be reviewed to establish areas where further enhancements of the peer-review process could be made. Moreover, it was recognized that due to differences in development between the inventory and impact components of LCA, it is likely that more definitive Data Quality Indicators (DQIs) may be established for the inventory analysis.

Research Needs Research needs identified during the workshop can be grouped into three general categories:

1. data quality assessment framework development;
2. data and data base development; and,
3. mathematical models advancement.

4.2.5 Code of Practice. Sesimbra, Portugal, March 31–April 3, 1993

Considering the four SETAC LCA workshops in Smugglers Notch (1990), Leiden (1991), Sandestin (1992) and Wintergreen (1992), it is noticeable that they formed a tiered process to culminate in the workshop of Sesimbra, Portugal, March 31–April 3, 1993.

The European and North American organizations of SETAC planned and conducted the LCA 'Code of Practice' Workshop in Sesimbra, Portugal. It was, after the workshop in Leiden, the second cooperative effort of the two SETAC LCA Advisory groups and the fifth technical workshop on LCA. Fifty experts from 13 countries were invited to define the LCA method and discuss its various possible applications.

Allan Astrup Jensen and Dennis Postlethwaite reported: "There were some 50 participants, deliberately selected to represent a wide range of views and opinions and to include a full representation of all interested parties—from Institutes, Governmental Bodies, Academia and Industry. The participants, and especially the organizing committee, worked very hard, such that the first report draft was circulated for comment by April 5th—quite a remarkable achievement. This draft was further debated and discussed by all participants, after which it was presented to a wider audience at Open Forums held in Europe and the United States. The final document, which incorporated significant points and amendments from these meetings, was published in the autumn of 1993. Because of legal implications in the US, the title had to be changed to a Guideline for Life Cycle Assessment with the subtitle: A Code of Practice [instead of 'Code of Conduct', as planned originally]. It was planned that an additional, longer and more methodological report should be made based on the Sesimbra discussions but, because of other pressures, this failed to materialize, much to the disappointment of several participants."

Objectives The 'Code of Practice' is not a standard for conducting LCAs. It provides guidance on process and methodological aspects of conducting LCAs reflecting the current situation, namely the status in 1993 which concerned the following issues:

1. LCA is a complex, multi-dimensional tool.
2. The LCA methodology has yet to be fully described. Of the three LCA components, only the Life Cycle Inventory Analysis has been well documented. The Life Cycle Impact Assessment methodology remains in development, and the Improvement Analysis has yet to be described conceptually.
3. The state-of-the art in 1993 is Life Cycle Inventory (LCI) and improvement of environmental performance based on LCI information.
4. New issues arise as practitioners continue to gain knowledge and experience on the application of LCA.

The 'Code of Practice' reaffirmed the findings of the Sandestin workshop, which, on its part, was an evaluation of the two workshops in Smugglers Notch (1990) and Leiden (1991). The reaffirmation concerned the technical framework with the components goal definition and scoping and inventory analysis as part of the framework, as well as the three-step impact assessment model, the impact categories.

The 'Code of Practice' was intended as guidance for all individuals who commission, carry-out, review, or use the results of an LCA, and should be used to enhance the quality, transparency, and credibility of such studies.

Peer Review Process For the first time, the 'Code of Practice' recommended and described a peer-review process to be a key feature in the advancement of LCAs, because it would enhance the scientific and technical quality of LCAs, help to focus study goals, data collection, and provide a critical screening of study conclusions, thereby enhancing study credibility.

The LCA peer review was relatively new and had not been fully tested and optimized. The 'Code of Practice' recommended that it should be more extensive than that traditionally used for the publication of research in scientific journals, for the following reasons:

- Because some LCA applications have regulatory and public-policy implications, a broad consultative approach is desirable in the review process to reach conclusions.
- Where proprietary information is used in LCA studies to reach conclusions that will be made public, protection of the proprietary information requires novel methods of peer review.
- The complexity of the data collection/definition and of the LCA process requires a more multidisciplinary peer review process than is required in most scientific studies.

Furthermore, the 'Code of Practice' recommended an interactive peer review process at various stages of the study for LCAs directed toward public audiences; this strategy can ensure the credibility of the study.

This interactive peer review ideally should be carried out in three phases:

Phase 1: at the beginning of the LCA to review the goals, scope, boundaries, and the data collection planned;
Phase 2: after initial data collection or modelling, to review the progress and offer advice or comments; and
Phase 3: at the final report stage, to review the adequacy of the study and the credibility of the conclusions.

While the three-phase peer review process is desirable, in some instances only a review of the final study report and supporting data may be possible ('a posteriori' peer review). The ISO 14040 (1997) did not make a recommendation for one form or the other. The 'Code of Practice' was more 'modern' in that respect and also described the review process in more detail (Klöpffer 1997, 2005, 2012).

Future Research Needs The identification of future research needs is a continuing process. One objective of the international SETAC LCA workshops was to identify such requirements. Initial needs were identified as follows:

- Data quality and database development;
- Methodology development (notably generic model development), allocation, energy accounting, and communication;
- Minimization of differences between methodologies;
- Gaining of public acceptance of the LCA concept and applications via communication and education;
- A code of conduct for undertaking LCAs.

ISO Standardization Developing international consensus on harmonized methods has been a goal of the SETAC LCA workshops. The 'Code of Practice' completed the harmonization process. Shortly after the workshop, during the autumn of 1993, the ISO standardization process was initiated.

4.3 SETAC LCA Workgroups from 1994 to 2000

In 1994, as a result of the SETAC LCA workshops, the LCA Advisory Group of SETAC and the LCA Steering committee of SETAC Europe[15] established individual work groups to address specific LCA issues.

'In this connection, it should be mentioned that the role of the steering/advisory LCA committees cannot be overestimated. The working groups are installed and supervised by these committees which also prepare the regular LCA sessions at the annual SETAC and SETAC Europe meetings, decide about publications, etc.' (Klöpffer 2006).

[15] LCA advisory group in the USA, LCA steering committee in Europe.

The following reports were developed and published by SETAC LCA workgroups before the year 2000:

- Towards a Methodology for Life Cycle Impact Assessment. SETAC Europe 1996 (Udo de Haes 1996)
- Simplifying LCA: Just a Cut? SETAC Europe 1997 (Christiansen 1997)
- Life Cycle Impact Assessment: The State-of-the-Art. SETAC (NA) 1997,1998 (Barnthouse et al. 1998)
- Streamlined Life Cycle Assessment SETAC (NA) 1999 (Todd and Curran 1999)

Thus the two topics, Life Cycle Impact Assessment (LCIA) and streamlined/simplified LCA, were addressed from SETAC Europe and SETAC North America respectively.

SETAC Europe Report: Towards a Methodology for Life Cycle Impact Assessment. September 1996

Report of the SETAC Life Cycle Assessment (LCA) Impact Assessment Workgroup, SETAC LCA Advisory Group: Life Cycle Impact Assessment—The State-of-the-Art 1998

The SETAC workshops in Sandestin, 1992 (Fava et al. 1993) and Sesimbra, 1993 (Consoli et al. 1993) revealed that, according to the current situation of that time, the component 'Life Cycle Impact Assessment' (LCIA) was still in development:

- Classification: Defined; requires further work
- Characterization: Conceptually defined and partly developed
- Valuation: Conceptually defined; different methods and approaches currently being used

Walter Klöpffer (Klöpffer 2006) commented: 'The methodology had to be discussed in a broader context and with a larger public. The occasion was the fourth SETAC Europe Annual Meeting in Brussels, in April 1994, during which a special symposium on LCIA was organized and published as a SETAC Europe Report'.

Moreover, the Life Cycle Impact Assessment workgroup of the LCA Advisory Group of SETAC and that of the LCA Steering Committee of SETAC Europe prepared two individual reports on LCIA.

The reports can be seen as complementary documents. They elaborate strong similarities but also a limited number of different positions. The latter 'will need continued discussion to be resolved as additional experience is gained'.

The SETAC Europe work group focused primarily on comparing methodologies and recommending methodological improvements, especially in the area of resources, normalization, and certain aspects of valuation. It refers to the results of the Leiden Workshop (1991).

The SETAC work group effort has resulted in a fundamental examination of the basic strengths and weaknesses of the Sandestin framework (1992). This has led to a number of significant insights, e.g., LCIA's being an indicator system. It has also seen the need to address uncertainty, to reach the understanding that LCIA indicators do not necessarily represent actual and significant differences between systems, and to integrate LCA with other analytical techniques and decision tools. This work group did not evaluate particular methodologies for any category.

Areas of Consensus

- Life cycle impact assessment deals with mass loadings as aggregated emissions and cannot assess actual impacts.
- Life cycle impact assessment is a simplified indicator approach.
- Risk assessments or Sandestin 'level 5' site-specific assessments are absolute in nature and cannot be conducted from relative LCA data after the functional unit and other inventory calculations.
- Spatial and temporal discontinuities exist between LCA and a number of environmental processes that affect the reliability and environmental representativeness of LCA information.
- There are interpretative implications of using LCA assumptions in representing environmental processes.

Issues Addressed by One Workgroup and not the Other The North American SETAC work group addressed:

- The identification and points of use of subjective judgments in classification and characterization.
- The integration of LCA results with other techniques and information.
- The need to address uncertainty in LCA and to distinguish differences between systems other than using point estimates based upon averages.
- The need to develop a better understanding of peer or critical review processes in regards to LCIA.

The SETAC Europe work group addressed:

- A concentrated effort to compare methodologies.
- A more detailed discussion of resource use methods.

Issues Where There May Be a Lack of Consensus The North American work group did not envision a default list of categories. One was apparently suggested by the European work group (see Table 2, page 15 in the SETAC Europe report). The North American work group was sceptical that a generic set of valuation weighting factors can be developed. However, the European work group was more optimistic.

SETAC Europe Report: Simplifying LCA: Just a Cut? Final report from the SETAC Europe LCA Screening and Streamlining Working Group. Editor: Kim Christiansen. May 1997

Streamlined Life Cycle Assessment: A Final Report from the SETAC North America Streamlined LCA Workgroup. Edited by: Joel Ann Todd and Mary Ann Curran. July 1999

In 1994 the LCA Steering Committee of SETAC Europe established a series of work groups, including the Workgroup Screening and Streamlining. In the same year, the SETAC North America workgroup on Streamlining LCA was initiated. Both groups concluded their multi-year efforts on the issue of Simplifying/Streamlining by a report in each case. The approaches of the reports were different.

The report of SETAC Europe discussed the methods for producing simplified procedures, commonly described as screening LCA studies, streamlined LCA studies and simplified LCA studies. The report of SETAC North America was more a description of carefully planning and stating an LCA's goal than it was about Streamlined LCA methodology. As can be seen from the two reports of the LCIA groups, the European position was more practical, while the USA point of view concerned the more theoretical, superordinate system.

Report of the LCA Steering Committee of SETAC Europe

Simplified LCA is an application of the LCA methodology for a comprehensive screening assessment. A simplified LCA should cover three steps which are iteratively interlinked:

1. Screening: identifying those parts of the system (life cycle) or of the elementary flows that are either important or have data gaps.
2. Simplifying: using the findings of Screening in order to focus further work on the important parts of the system or of the elementary flows.
3. Assessing reliability: checking that simplification does not significantly reduce the reliability of the overall result.

Simplifying methods can reduce the complexity of an LCA and so reduce the cost, time and effort required, by exclusion of certain life cycle stages, system inputs or outputs or impact categories, or use of generic data modules for the system under study.

Report Organization

- Chapter 1 introduces the issues of screening and simplifying in LCA.
- Chapter 2 lists the definitions on screening and simplifying concepts used in this report.
- Chapter 3 continues the introduction by developing the framework for the three-step process of simplifying an LCA and discussing each in detail.
- Chapter 4 discusses the reporting of a simplified LCA.
- Chapter 5 presents a series of examples.
- Chapter 6 presents the conclusions and recommendations from the workgroup for further work in the area of simplifying LCA.

Application of Methods: Simplification The goal and scope definition should not be simplified itself. Goal definition, as such, cannot be reduced to defining the goal of a simplified LCA as a goal definition. The scope definition, similarly, cannot be minimized, but the borders of the product system can be set to give a simplified picture of the product system. It is, however, important to remember that the first step of this procedure, screening, should cover the product system from cradle-to-grave (i.e., be comprehensive). Regarding available data, both the main processes and ancillary processes are significant to the product system.

The life cycle inventory analysis phase offers the greatest scope for simplification. It primarily involves the use of readily available data representing the product system at a generic and not system-specific level.

In the life cycle impact assessment phase, many of the classifications, characterizations and weighting methods available are already simplified versions of much more detailed environmental assessment approaches. Yet, the selection of impact categories, the impact data applied, etc. can sometimes be narrowed down without losing the overall quality and reliability of the LCA study. Simplifying the life cycle inventory analysis will imply a simplification of the life cycle impact assessment.

The life cycle interpretation phase can only be simplified with great caution. Interpretation as described by ISO/CD 14043 is a simplification procedure in itself, for the benefit of a more understandable and assessable result of the life cycle inventory analysis and/or the life cycle impact assessment. LCA expertise, as well as product system expertise, is still needed in the interpretation.

Report of the SETAC North America Workgroup on Streamlining LCA

Streamlined LCA: Identification of elements of an LCA that can be omitted or where surrogate or generic data can be used without significantly affecting the accuracy of the results.

Streamlining LCA is a practice to make a detailed/full LCA more manageable. Streamlining LCA can be achieved in a number of ways, including:

- Limiting the scope in terms of time, cost, data, analytical approach: for example, eliminating life cycle phases deemed not significant, or processes with negligible effect on the environment;
- Use of qualitative information;
- Removal of upstream and/or downstream components;
- Use of specific impact category.

Streamlining is an inherent part of any LCA. The key is to link the streamlining activities closely with the goal and scope definition process. That is, streamlining is a routine element of defining the boundaries and data needs of a study and is not in itself a different approach or methodology for LCA. In other words, 'full-scale' LCA and 'streamlined' LCA are not two separate approaches but rather two points on a continuum.

Report Organization The report consists of 4 chapters:

- Chapter 1 is a short introduction to LCA and the issues surrounding simplifying the process.
- Chapter 2 describes the important role of the goal-and-scope definition process in streamlining decisions.
- Chapter 3 describes approaches to streamlining.
- Chapter 4 offers some concluding thoughts on streamlined LCA methods.

The purpose of this report is to:

- redefine streamlining as an inherent part of any LCA approach that involves deciding what is and what is not to be included in a study;
- emphasize that streamlining steps must be consistent with the original study goals and anticipated uses;

- describe various ways that streamlining LCA has been attempted and investigated and the possible implications in different decision-making contexts; and
- provide recommendations on how the goal-and-scope definition process can be used to design and streamline an LCA study.

The workgroup and thus the report have also benefitted from the work of the SETAC Europe Workgroup on Streamlining and their report 'Simplifying LCA: Just a Cut?'

Within the workgroup, however, consensus has not been reached on the exact methods and procedures that can be used in a streamlined LCA or on appropriate uses of a streamlined LCA.

4.4 SETAC LCA Workshops and Initiatives up from 1999

In 1994, the SETAC Europe LCA steering committee and the SETAC (North America) LCA advisory group established, amongst others, the workgroups on Life Cycle Impact Assessment and Simplified/Streamlined LCA. These workgroups published four reports (see Sect. 4.3 'SETAC LCA workgroups from 1994 to 2000').

Walter Klöpffer (2006) focused in his article on the workgroups on Life Cycle Impact Assessment (LCIA), due to the complexity of the issue and the difficulties arising from the somewhat different approaches of the LCIA workgroups (SETAC Europe and SETAC North America); they are defined in the Sect. 4.3 'SETAC LCA workgroups from 1994 to 2000').

According to Klöpffer it transpired that the European efforts towards a unified LCIA methodology did not get acceptance by the global LCA community. Sessions on LCIA were organized at the annual meetings following the Brussels (4th) annual meeting in 1994 (Udo de Haes et al. 1994).

The LCIA issue was followed up in a group chaired by Helias Udo de Haes. Subgroups had to be formed to handle the broad topic. Within one year, the framework paper was published in Int J Life Cycle Assess (Udo de Haes et al. 1999a, b) in the form of two reports. They constitute a basis for the identification of best available practice concerning impact categories and characterization factors for Life Cycle Impact Assessment. The reports are the result of the first working phase of the second workgroup on Life Cycle Impact Assessment of SETAC Europe. In this workgroup members from other divisions of SETAC participated, in particular from the USA and from Japan.

Thus, the framework paper was ready to be discussed during the SETAC Europe 1999 annual meeting in Leipzig. A discussion with SETAC members from North America took place as well, even publicly (Owens 1998, 1999). The framework paper was finally ready for review in the end of 2001 and published by SETAC Press 2002 in the award-winning book 'Life Cycle Impact Assessment: Striving Towards Best Practice' (Udo de Haes et al. 2002a).

During the period from 1994 to 2000 (first wave of workshops, see section *'SETAC LCA workgroups from 1994 to 2000'*), another workshop took place and was published by SETAC:

4.4.1 Application of Life Cycle Assessment to Public Policy, August 14–19, 1995, Wintergreen, VA, USA

Public Policy Applications of Life Cycle Assessment. Proceedings from the Workshop on 'Application of Life Cycle Assessment to Public Policy', 14–19 August 1995, Wintergreen, Virginia, USA. Edited by David T. Allen, Frank J. Consoli, Gary A. Davis, James A. Fava, John L. Warren.

The life cycle concept is a powerful systems approach for thinking about technology from a 'cradle-to-grave' perspective. Life cycle assessment (LCA) is one analytical tool for implementing life cycle concepts. Use of life cycle concepts and tools can link scientific, technological, and policy-making communities in an overall effort to find an appropriate balance between economic, environmental, and energy considerations.

Public policies are actions, decisions, statements, mandates, orders, or guidance taken by governmental entities that affect other governmental entities, non-governmental entities, the public, and private interests. Public policies are shaped and constrained by many interests within and outside government, existing policies, legal and societal norms, and institutional arrangements.

Governmental entities include all regulatory and non-regulatory institutions (e.g., programs, agencies, departments) at local, state, regional, provincial, federal, and international levels, across executive, legislative, and judicial branches.

The shift toward more integrative public policy tools is occurring at all levels of government and around the world.

The life cycle concept may improve the public policy process by providing information to decision-makers in a comprehensive manner. Public policy decisions, however, are extremely varied. They range from the implementation of narrow mandates to the development of broad policy statements and involve very different institutions, from local planning departments to federal agencies and the Executive Office of the President. Consequently, the application of the life cycle concept to public policy will involve a broad range of depth, breadth, and rigor.

For life cycle concepts to be widely applied in policy settings, the results must be understandable, transparent, and accessible to all stakeholders. This level of understanding will require an aggressive education and information-dissemination effort.

The objectives of the workshop were to:

- define the public policy arenas in which LCA could inform decision-making;
- develop specific guidance for the use of LCA in public policy;
- specify a framework and approach for LCA use in public policy decisions; and,
- determine future research needs in the application of LCA to public policy-making.

Approximately 40 internationally recognized experts in LCA and the application of LCA to public policy-making were organized into five workgroups. One working group was charged with outlining a framework for the application of lifecycle assessment to public policy. Their report is contained in Chapter 2. The remaining 4 groups considered specific application arenas. Chapter 3 reports on the use of LCA in environmental labeling initiatives. Chapter 4 examines the use of LCA in governmental acquisition and procurement. Chapter 5 considers the use of LCA in analyzing regulations and setting policy, and Chapter 6 probes the use of LCA in identifying environmental technologies.

4.4.2 A Second Wave of LCA Workshops

During the Bordeaux (8th) annual meeting in 1998, a second wave of workgroups was started with a planned duration of three years. Within the SETAC Europe workgroups, several other reports were published:

- Code of Life Cycle Inventory Practice (Beaufort-Langeveld et al. 2003)
- Life cycle management (LCM) (Hunkeler et al. 2004)
- The working environment in LCA (Poulsen et al. 2004)
- Scenarios in Life Cycle Assessment (Rebitzer and Ekvall 2004)
- Life Cycle Assessment in Building and Construction (Kotaji et al. 2003)

One prerequisite for the cooperation between UNEP (United Nations Environment Program) and SETAC, was the transformation of SETAC into a truly global organization in the late 1990s. Another reason was UNEP's need for implementing sustainable development, proclaimed as the most important goal of humankind in Rio de Janeiro 1992[16] and confirmed in Johannesburg in 2002[17]. Sustainability is based on methods derived from life cycle thinking (Klöpffer 2003), with LCA as the core element. Thus, a co-operation between SETAC and UNEP's Production and Consumption Branch (Paris), was logical and promising. The cooperation between UNEP and SETAC was officially launched on April 28 2002, in Prague (Töpfer 2002).

The key people from SETAC in the negotiations were Jim Fava (Fava 2002), Helias Udo de Haes and Olivier Jolliet (Udo de Haes et al. 2002b). In 2003, The International Journal of Life Cycle Assessment became the associated journal of the Initiative (De Larderel and Fava 2003).

The UNEP/SETAC Life Cycle Initiative is an achievement in significant part due to SETAC. With this success, however, several areas (e.g., LCI, LCIA, LCM, traditionally covered by SETAC and SETAC Europe) were now primarily addressed by the Initiative. There is continuing efforts to ensure complementary programs within

[16] United Nations Conference of Environment and Development (UNCED), Rio de Janeiro, June 1992.

[17] World Summit on Sustainable Development, Johannesburg, September 2002.

UNEP/SETAC LC Initiative and SETAC building off the skills, and skills of each group to advance the development and application of life cycle approaches globally.

Klöpffer (2006) scrutinized the future of LCA in SETAC and recommended that SETAC as the scientific arm of the Initiative should continue to take the leadership in LCA. In this connection he pointed to two workgroups "which may be especially promising for further 'cutting edge' activities":

- Life Cycle Costing
- Input-Output and Hybrid Life Cycle Assessment

Another important item in LCA—and a deficiency—would be the inability of LCIA to incorporate non-chemical impacts to ecosystems, e.g., invasive species and certain biotechnologically modified organisms.

Further, 'Sustainable consumption' would turn out to be a field of considerable interest.

Klöpffer (2006) recommends that the SETAC work groups, enacted and surveyed by the LCA advisory/steering committees, should continue to play a major role in the development of life cycle based assessment and management methods.

5 SETAC and the International Organization for Standardization[18]

The International Organization for Standardization (ISO) is the world's largest developer of voluntary International Standards[19]. In the fall of 1993, when ISO was questioning the need for an international life cycle assessment standard, they nominated a small group consisting of SETAC LCA experts and others to develop recommendations to consider whether the LCA standardization should be pursued. The Strategic Advisory Group on the Environment (SAGE) chaired by Jim Fava[20], brought together the international experts on LCA involved in SETAC's LCA Groups and other international experts, to develop the recommendation.

The SAGE recommended that standards should be written on General Principles and the Life Cycle Inventory Phase of LCAs, but suggested that Impact Assessment and Interpretation phases of LCA were not yet developed enough to be included in ISO LCA standards. The two recommended standards were suggested as they already had international methodologies in place and required only harmonization as opposed to development. However, during the first ISO meeting to discuss the LCA standards, overwhelming interest from many countries led to the final decision

[18] See this volume, Chapt. 5 'The international standards as constitution of LCA: the ISO 14040 series and its offspring' by Matthias Finkbeiner.

[19] http://www.iso.org/iso/home/about.htm.

[20] Jim Fava, Chair, at the time Vice President, WESTON Solutions.

that four standards[21] (ISO 14040 to 14044) would be written so that more countries could be involved. While standards based on General Principles and the Life Cycle Inventory Phase of LCAs were developed quickly, the additional two standards, on Impact Assessment and Interpretation were slow to develop as they created the need to develop new methodologies rather than just harmonizing existing ideas. Today, these four original LCA standards have been combined into two.

6 On-Going SETAC Activities

In the years when many North American SETAC members were involved in developing the ISO LCA standards, additional SETAC activities were occurring outside of North America. In Europe, professors and students were developing LCA methodology, leaving other SETAC members to work on both SETAC activities and ISO standard development. SETAC Europe continued developing working groups and addressing various international environmental concerns.

6.1 Global Advisory Groups

The Global Coordinating Group (GCG) was formed in 2011 as a mechanism for communication between the regional Advisory and Steering Groups in North America and Europe to allow the other Geographic Units to have representation in global SETAC LCA affairs. The GCG became the point of membership for all members in the SETAC LCA Community with the switch to the new membership platform.

The two geographic unit level LCA advisory and steering groups were among the earliest established in SETAC. At the time, distinct interests and centres of activity made it appropriate to have separate groups for the two regions. As time went on, LCA became a more global practice and the interests of regional groups more overlapping and intersecting. In addition, LCA practitioners and interested individuals from Latin America, Asia-Pacific, and Africa had no direct voice in the advisory and steering group governance structure existing within SETAC.

The mission of the LCA Global Coordinating Group within SETAC is to encourage and coordinate regional Advisory Group efforts to advance the science, practice, and application of LCAs, and to ensure that a global perspective is maintained toward the achievement of LCA Groups objectives. To accomplish this mission, the Group serves as a focal point to provide a harmonizing forum for the identification, resolution, and communication of issues and activities regarding LCA across geographic units. Further, it facilitates, coordinates, and provides guidance for the development and implementation of LCA methodology and practice in close coop-

[21] The four standards: General Principles of LCA; Life Cycle Inventory Phases of LCA; Impact Assessment LCA; and Interpretation of LCA.

eration with the LCA Groups of SETAC in various regions. As such, it serves as a point of liaison between the SETAC LCA Groups and the SETAC governance role as co-Chair of the International Life Cycle Board of the UNEP/SETAC Life Cycle Initiative.

The Global Coordinating Group's website[22] contains a collection of information on various SETAC LCA topics and provides links to relevant resources.

7 UNEP/SETAC Life Cycle Initiative[23]

In 2002, the United Nations Environment Programme (UNEP), the Society of Environmental Toxicology and Chemistry (SETAC) and partners from governments, academia, civil society, business and industry joined forces to promote life cycle approaches worldwide. This was done to increase resource-efficiency and to accelerate a transition towards more sustainable consumption and production patterns. Sustainable development objectives and a company's bottom line come together in the important topic of assessing and managing the life cycle of processes, materials, products and services.

After the publication of the ISO 14040 standard dealing with LCA (ISO 14040:1997), UNEP and SETAC became aware of the need for dissemination and implementation. They jointly began to engage more partners to work on the articulation of science-based existing efforts around life cycle thinking and established the UNEP/SETAC Life Cycle Initiative (Life Cycle Initiative).

This life cycle partnership for a more sustainable world between UNEP, SETAC and public/private sector partners has the overall objective of promoting, assisting and supporting the application of life cycle thinking and life cycle approaches. This includes life cycle management, life cycle assessment, carbon footprinting and water footprinting, by governments as well as companies and their suppliers, customers and other value-chain partners worldwide. The final purpose is furthering sustainable innovation and global use of more sustainable products.

The Life Cycle Initiative's activities have been carried out in three phases, in which around 2000 members of the global life cycle community have been actively involved.

The first phase (2002–2007) focused on establishing the Life Cycle Initiative as a global focal point of life cycle-related knowledge and activities and building an expert community of practitioners. Activities to move the Life Cycle agenda forward concentrated on three important fields of work:

1. Life Cycle Management (LCM);
2. Life Cycle Inventory (LCI); and,

[22] http://www.setac.org/group/AGLCA.

[23] See also Chap. 6 'The UNEP/SETAC Life Cycle Initiative' by Guido Sonnemann and Sonia Valdivia.

3. Life Cycle Impact Assessment (LCIA) as well as the crosscutting area of social impacts along the life cycle.

At the end of the first phase, a process was started to help create regional and national life cycle networks, particularly in developing countries, to support capability development. Due to the important personal engagement of Greg Norris, today faculty of the Harvard School of Public Health, it was possible to get life cycle networks in Africa and Latin America initiated.

Phase 2 activities (2007–2012) saw the Life Cycle Initiative evolve to be more participative with regard to stakeholders, encouraging more involvement from key actors at the global level. The goal was to achieve common understanding and agreement on tools and strategies being developed. The main outcomes of Phase 2 were accomplished through close collaboration with crucial stakeholders in the field.

In both Phase 1 and 2, the Life Cycle Initiative was able to provide support in the application of sustainability-driven life cycle approaches based on lessons learned from leading organizations by its capacity of engaging with world class experts and practitioners working in product policy, management and development.

The Life Cycle Initiative started Phase 3 in 2012 with a mission to 'enable the global use of credible life cycle knowledge for more sustainable societies'. Its overarching goal is to 'facilitate the generation and uptake of science-based life cycle approaches and information for products and organization by business, government and civil society practice worldwide as a basis for sustainable consumption and production'. Activities in Phase 3 will focus on creating the enabling conditions to: (a) enhance the global consensus and relevance of existing and emerging life cycle methodologies and data management; (b) expand capabilities worldwide and make life cycle approaches operational for organizations; and, (c) communicate current life cycle knowledge and be the global voice of the life cycle community to influence and partner with stakeholders.

8 SETAC's Role in Advancing the Use of LCA in the Building Sector

In the fall of 2004, there was a unique opportunity for two leading organizations to come together and begin a dialogue on the use of life cycle approaches within the building sector. SETAC and the U.S. Green Building Council's (USGBC) Green Build forum both were meeting in Portland, OR, USA, on back to back weeks. This opportunity was well timed, as both the building and construction sector and SETAC had progressed to a point of being ready for collaboration.

At the time, the building and construction sector was beginning to recognize that the impact of building construction and operations was significant and it was obvious that a systems and life cycle approach would support progress in reducing the footprint associated with building material. While the life cycle community had been able to advance the life cycle methodology, the practical applications to specific sectors was one of the next steps in life cycle advancement. The building sector was a logical sector-specific application of life cycle approaches.

SETAC, USGBC and the UNEP/SETAC Life Cycle Initiative[24] organized a one-day forum in the fall of 2004 entitled "Advancing the use of life cycle approaches by building decision makers." The purpose of the forum was to provide an opportunity to exchange information on LCA and green buildings programs. During these discussions, SETAC and the UNEP/SETAC Life Cycle Initiative provided the North American Green Building community (USGBC, manufacturers, architects, city/state/federal government, consultants, NGOs, academics) with a chance to come up to speed on the current state of LCA applied to construction in Europe, Australia, Latin America and Asia Pacific.

At the same time, the North-American green building community provided members of the life cycle community from Australia, Europe and other regions with an understanding of current events and trends in application and policies related to life cycle approaches to green construction in North America. Additionally, SETAC and the UNEP/SETAC Life Cycle Initiative provided the building community with a description of other international programs and forums. These included the UNEP/SETAC Life Cycle Initiative, the ISO process on updating LCA standards, and the ISO process to develop a framework for assessment of environmental performance of buildings, all of which were shaping LCA and its application to buildings.

These groundbreaking discussions laid the solid foundation necessary for the continued exploration of the application of LCA within the building and construction sector. There was a solid interest in advancing LCA through collaboration within the building and construction sector generally and with USGBC specifically. While details on how to expand the use of LCA within the building and construction sector were not the focus of the workshop, the group felt they were at a critical point related to the use of LCA in advancing Green Building decision making. In the view of one attendee, 'a 15 year old tool has finally found a purpose.'

The ISO 14040 family of LCA standards should be used as a starting point for further development of LCA methodology within the building sector application. The use of LCA tools must also ensure that barriers to trade are not developed. It was also strongly pointed out that, while LCA may be a useful tool to improve green building decision making, it is not the only tool. Other tools and information will be needed to improve green buildings.

Two issues, among others, that surfaced during conversations between the two groups included the need for further examination of a 'functional unit' for buildings, and the pros and cons of performance and/or continuous improvement based

[24] The Advisory Committee for the Forum: Jim Fava, Chair, Managing Director of Five Winds International, and Vice-chair for the UNEP/SETAC International Life Cycle Panel; Deborah Dunning, President, International Design Center for the Environment; Pamela Horner, Sylvania and IESNA; Gregory A. Norris, Sylvatica, and Programme Manager for the UNEP/SETAC Life-Cycle Initiative program; Bob Peoples, Carpet & Rug Institute and CARE; Guido Sonnemann, UNEP/SETAC Life Cycle Initiative Secretariat; and, Wayne Trusty, President, Athena Sustainable Materials Institute.

approaches to using LCA. LCA can now be used at two levels, at the level of the building as a whole and at the level of building materials or products. Experiences obtained so far indicate that the latter is easier than the former, although applications at the building level can also produce useful results and are advancing.

The building and materials market has radically changed over the last ten years or so. In the United States, the Green Building Council (USGBC) developed its LEED® system. In the UK and Germany, similar programs have created the foundation for market transformation. Today there are dozens of Green Building Councils around the world. While the rating systems are not perfect, they have created the capacity to allow architects, designers, and building commissioners to integrate sustainability into the building and construction sector.

9 Future Role of SETAC

As SETAC works to advance the understanding and use of LCA, it will continue to ensure that science is kept in the forefront of LCA development. By doing so, SETAC will continue to help LCA remain credible and trusted. This overarching focus, along with its emphasis on balanced engagement among academia, business and government, will remain in all of SETAC's activities, including those that are highlighted in the following sections[25].

9.1 Expanding the Use of LCA

Although SETAC has advanced the development and implementation of LCA in Europe and North America, it has yet to grow these activities in other geographic areas. Particularly in Africa and South America, SETAC is in the early stages of incorporating LCA into their regional meetings. Connecting the right people is a necessary part of developing LCA in an area currently unfamiliar with it. For example, SETAC is in the early stages of connecting toxicologists with individuals focusing on LCA so that a team with local knowledge can be formed.

9.2 LCA Case Studies

SETAC Europe's Case Study Symposium, which is entering its 19th year, is a forum for LCA professionals to share case studies with an international audience. Although the Case Study Symposium is currently an opportunity to share case study results, there has been interest to evolve the symposium into a platform to assemble

[25] The following sections are based on conversations with Bruce Vigon, Scientific Affairs Manager at SETAC.

and critically examine those results as well. This critical examination will help the LCA community to evaluate the science behind LCA and ensure that LCA remains reputable. As well, it should allow students and early career professionals to share their work.

Rick Wenning, the Editor-in-Chief of Integrated Environmental Assessment and Management (IEAM)[26], has expressed a willingness to increase the presence of LCA in the SETAC Journals. The organizers of the 2012 Case Study Symposium in Copenhagen recently assembled a special issue of the journal that consisted of the top ten symposium papers. Because the issue underwent peer-review as part of the normal publication process, the collection of papers has professional acceptance and is more highly regarded. This, in turn, reflects favorably on the symposium activity. Using this special issue of IEAM as a guide, SETAC can enhance its visibility by publishing peer-reviewed collections from future events.

9.3 Additional Pellston Workshops

The Pellston Workshop format, though proven to be successful over decades, including the purpose of further developing LCA, should be used only when circumstances such as the need for global consensus dictate it. Considering the Pellston Workshop requires such a rigorous effort, one of the other workshop formats that SETAC has developed over the years can often suffice to meet the scientific and publication objectives and be less resource intensive.

In order to require a Pellston Workshop, the workshop topic being covered needs to be worthy of the Pellston brand. This means that the issue needs to be one that is controversial, lacking in the needed consensus, and has a significant research or scientific question that only the rigor of the Pellston workshop is capable of addressing. For example, topics that wouldn't be worthy of a Pellston workshop, would be those that are regional or country-specific, exploratory in nature, or those that have a lack of international interest.

The Pellston format should be reserved for workshops that will provide an impactful, consequential publication. The intense nature of the workshop results in conclusions being reached. However, although this outcome is enticing, the Pellston Workshop requires a twelve to fifteen month commitment that cannot be taken lightly.

Outside of its pre-existing workshops, SETAC has developed three new types of events. The North American Focused-Topic Meetings, and Special Science Symposiums in Europe, are common in terms of the topics they cover and the outputs that they produce. The purpose of both meetings is to disseminate and exchange current interdisciplinary information on a specific environmental topic.

---

[26] http://onlinelibrary.wiley.com/journal/10.1002/(ISSN)1551-3793.

The third new event type, the Invited Conference, is based on the Gordon Research Conference[27] model and is invitation only. Because the Invited Conferences have less strict publication requirements, the conversations are more open.

9.4 *On-Going Effort with the UNEP/SETAC Life Cycle Initiative*

SETAC's role within the UNEP/SETAC Life Cycle Initiative has always been to provide substantive scientific expertise as well as the means to deliver this expertise through forums such as the workshops, meetings and symposia described above. Since this role is so significant, it is important that SETAC members continue to be involved in new issues and have the opportunity to join working teams that are addressing new issues. Recently, the organizers of the Initiative identified topics in impact assessment where science is a necessary and high level component.

Although many topics can be tackled by the UNEP/SETAC Life Cycle Initiative, there is still a difference in emphasis on contributions from UNEP, SETAC, or the UNEP/SETAC Life Cycle Initiative. Areas where science still needs to be developed and applied should be major contribution areas by SETAC. For example, it is problematic to address new technologies when there is no existing facility operating to provide actual data within an LCA. How to model new technology that is likely to impact the environment differently than current technology while maintaining comparability and similar uncertainty to datasets from existing well commercialized technologies is a scientific challenge well suited to SETAC member expertise. These emerging or developmental approaches could ultimately be brought into the endorsed set of methods under the Life Cycle Initiative.

9.5 *Impact Assessment Advancement*

SETAC members have methods and tools that were developed for environmental risk assessment that can be put towards impact assessment advancement. Specifically, SETAC is aligned to be involved in promoting the evolution of a framework and inventory that can support the science of increasing relevant impact assessment methodology. However, not all impact categories are necessarily going to follow the existing LCA methodology. It will be vital for SETAC to bring together a group of impact assessment experts to evaluate the social and environmental impacts that can be measured within LCA. The group would determine what can be done within impact assessment. While some aspects of impact assessment won't be measureable within the logical structure of LCA, some pieces are already capable of being measured in an LCA framework.

---

[27] http://www.grc.org/.

9.6 *Alternative Assessments*

SETAC is beginning to expand and alternatively assess materials and chemicals. SETAC's assessment will evolve past toxicological assessment and evaluating risks on the environment and human health to include broader life cycle impacts and life cycle stages. SETAC's unique membership can bring together risk assessment and LCA experts to determine if one chemical or material is more environmentally sound than another.

9.7 *LCA in Developing Countries*

SETAC's role in LCA development within North America and Europe is much different than its role in developing countries. In areas where LCA is a well-established and practiced tool, SETAC needs to begin to look to specific areas to build and maintain fundamental science as the scope and capability of LCA is expanded and refined. In developing countries, SETAC's role mission for education is more relevant. UNEP uses a formulaic approach to evolving LCA in developing countries (i.e., by gathering local representatives and developing intergovernmental forums). SETAC's role should be to use the LCA community to help develop LCA expertise by hosting learning sessions, webinars and developing education tools for developing countries that are advancing in using LCA. Rather than SETAC's educational focus competing with UNEP/SETAC's structural focus, the two complement each other.

As SETAC is beginning to educate developing countries on LCA, it is important to ask what type of education is relevant. SETAC proposes to develop courses concerning LCA with the aim to bring together local toxicology experts that are able to contribute to local LCA development, what is meant by LCA, why is it important, and what they can do to start building their own internal capacities. This will include a focus on how to create a connection between the regional/local situation and existing LCA frameworks and methods.

Appendix—Glossary

Global Coordinating Group (GCG)	The Global Coordinating Group (GCG) was formed in 2011 as a mechanism for communication between the regional Advisory and Steering Groups in Europe and North America and to allow the other Geographic Units to have representation in global SETAC LCA affairs

International Organization for Standardization (ISO)	After the LCA harmonization by SETAC, shortly after the workshop in Sesimbra in 1993, the LCA ISO standardization process was initiated
LCA in developing countries	SETAC's role should be to use the LCA community to help develop LCA expertise by hosting learning sessions, webinars and developing education tools for developing countries that are advancing in using LCA
LCA in the building sector	LCA can now be used at two levels, at the level of the building as a whole and at the level of building materials or products
Pellston workshops	Pellston Workshops, named after the location of the first workshops (University of Michigan Field Station, Pellston, MI, USA). The goal of a Pellston Workshop is to promote advancement in the resolution of truly cutting-edge technical and policy issues in environmental science, while enhancing strategies of science and philosophy
SETAC LCA groups	The SETAC LCA European group is named 'LCA Steering Committee' The SETAC LCA North American group is named 'LCA Advisory Group'
Technical workshops	SETAC supports the convening of technical workshops to bring together experts to discuss and resolve timely technical, scientific or policy issues related to environmental science. SETAC's level of support can range from simply providing an endorsement (e.g., non-exclusive license to use SETAC name or logo for promotional purposes) to providing full technical and scientific support, as long as basic principles are met
UNEP/SETAC Life Cycle Initiative	In 2002, the United Nations Environment Programme (UNEP), the Society of Environmental Toxicology and Chemistry (SETAC) and partners from governments, academia, civil society, business and industry joined forces to promote life cycle approaches worldwide. This was done to increase resource-efficiency and to accelerate a transition towards more sustainable consumption and production patterns. Sustainable development objectives and a company's bottom line come together in the important topic of assessing and managing the life cycle of processes, materials, products and services
Work groups Life Cycle Impact Assessment	The Life Cycle Impact Assessment workgroup of the LCA Advisory Group of SETAC and that of the LCA Steering Committee of SETAC Europe prepared two individual reports on LCIA. They can be seen as complementary documents. The reports elaborate strong similarities but also a limited number of different positions

	SETAC Europe Report: Towards a Methodology for Life Cycle Impact Assessment. September 1996
	Report of the SETAC Life Cycle Assessment (LCA) Impact Assessment Workgroup, SETAC LCA Advisory Group: Life Cycle Impact Assessment—The State-of-the-Art 1998
Work groups Simplified/Streamlined LCA	In 1994 the LCA Steering Committee of SETAC Europe established the Workgroup Screening and Streamlining. In the same year, the SETAC North America workgroup on Streamlining LCA was initiated. Both groups concluded their multi-year efforts on the issue of Simplifying/Streamlining by a report in each case. The approaches of the reports are different
	SETAC Europe Report: Simplifying LCA: Just a Cut? Final report from the SETAC Europe LCA Screening and Streamlining Working Group. Editor: Kim Christiansen. May 1997
	The report of SETAC Europe discusses the methods for producing simplified procedures, commonly described as screening LCA studies, streamlined LCA studies and simplified LCA studies
	Streamlined Life Cycle Assessment: A Final Report from the SETAC North America Streamlined LCA Workgroup. Edited by: Joel Ann Todd and Mary Ann Curran. July 1999
	The report of SETAC North America is more a description of carefully planning and stating an LCA's goal than it is about Streamlined LCA methodology
Workshop Leiden	'Life-Cycle Assessment—Inventory, Classification, Valuation, Data Bases'. December 2–3, 1991, Leiden, The Netherlands
	It was concluded that the term LCA can best be interpreted as 'life cycle assessment' instead of 'life cycle analysis'
Workshop Sandestin	'A conceptual framework for life cycle impact assessment'. February 1–7, 1992, Sandestin, Florida
	The aim of this workshop was to develop a consensus on the state of the practice and research needs for conducting life cycle impact assessments. The workshop reaffirmed the value of the three-component model for LCAs developed at the Smugglers Notch workshop in 1990. Also, building on the results of the Leiden workshop in 1991, a goal definition and scoping component was incorporated as an additional step
Workshop Sesimbra	'Code of Practice'. Sesimbra, Portugal, March 31 to April 3, 1993

	The European and North American organizations of SETAC planned and conducted the LCA 'Code of Practice' Workshop The 'Code of Practice' was intended as guidance for all individuals who commission, carry-out, review, or use the results of an LCA, and should be used to enhance the quality, transparency, and credibility of such studies Shortly after this workshop, the LCA ISO standardization process was initiated
Workshop Smugglers Notch	'A technical framework for life cycle assessment'. August 18–23, 1990, Smugglers Notch, Vermont The workshop was to develop a framework and consensus on the current state of LCA and research needs for conducting life cycle assessments. Although life cycle assessments have been used, in one form or another, before the name was coined, this workshop report is the first document which presented the name of the method
Workshop Wintergreen	'Data quality: a conceptual framework'. October 4–9, 1992, in Wintergreen, Virginia The workshop provided a strong statement that data quality assessment is an integral part of LCA

References

Barnthouse L, Fava JA, Humphreys K, Hunt R, Laibson L, Noesen S, Norris G, Owens J, Todd J, Vigon B, Weitz K, Young J (eds) (1998) Life-cycle impact assessment—the state-of-the-art. Report of the SETAC life-cycle assessment (LCA) impact assessment workgroup, SETAC LCA advisory group. 2nd edn. SETAC Pensacola, FL, USA

Beaufort-Langeveld ASH de, Bretz R, van Hoof G, Hischier M, Jean P, Tanner T, Huijbregts M (eds) (2003) Code of life-cycle inventory practice. (book + CD) ISBN 1-880611-58-9. SETAC Press, Pensacola, FL, USA

Christiansen K (ed) (1997, May) Simplifying LCA—just a cut? Final report from the SETAC Europe LCA Screening and Streamlining Working Group. SETAC Europe. Brussels, Belgium

Consoli F, Allen D, Boustead I, Fava J, Franklin W, Jensen AA, de Oude N, Parrish R, Perriman R, Postlethwaite D, Quay B, Séguin J, Vigon B (eds) (1993) Guidelines for life-cycle assessment: a 'Code of Practice'. Edn 1. From the SETAC workshop held at Sesimbra, Portugal, 31 March–3 April 1993. SETAC Pensacola, FL, USA

de Larderel JA, Fava J (2003) Int J LCA—Associated Journal of the UNEP/SETAC Life Cycle Initiative (Editorial). Int J Life Cycle Assess 8(3):117

Ekvall T (2005) SETAC summaries. J Clean Prod 13(13–14):1351–1358

Fava JA (2002) Life cycle initiative—a joint UNEP/SETAC partnership to advance the life-cycle economy. Int J Life Cycle Assess 7(4):196–198

Fava JA, Denison R, Jones B, Curran MA, Vigon B, Selke S, Barnum J (eds) (1991) SETAC workshop report: a technical framework for life-cycle assessment. August 18–23, 1990, Smugglers Notch, Vermont. SETAC, Washington, DC, USA

Fava J, Consoli FJ, Denison R, Dickson K, Mohin T, Vigon B (eds) (1993, Mar) A conceptual framework for life-cycle impact assessment. Workshop report. SETAC and SETAC foundation for environmental education. Sandestin, Florida, February 1–7, 1992. Published by SETAC and SETAC foundation for environmental education, Pensacola, FL, USA
Fava J, Jensen AA, Lindfors L-G, Pomper S, De Smet B, Warren J, Vigon B (eds) (1994, June) Life-cycle assessment data quality: a conceptual framework. Workshop report. SETAC and SETAC foundation for environmental education. Wintergreen, Virginia, October 1992. Published by SETAC and SETAC foundation for environmental education, Pensacola, FL, USA
Finkbeiner M, Schau EM, Lehmann A, Traverso M (2010) Towards life cycle sustainability assessment. Sustainability 2(10):3309–3322. doi:http://dx.doi.org/10.3390/su2103309
Hunkeler D, Saur K, Rebitzer G, Finkbeiner M, Schmidt WP, Jensen AA, Stranddorf H, Christiansen K (2004) Life-cycle management. ISBN 1-880611-77-5. SETAC Press, Pensacola
Hunt R, Franklin WE (1996) LCA—how it came about. Personal reflections on the origin and the development of LCA in the USA. Int J Life Cycle Assess 1(1):4–7
Klöpffer W (1997) Peer (expert) review in LCA according to SETAC and ISO 14040—theory and practice. Int J Life Cycle Assess 2(4):183–184
Klöpffer W (2003) Life-cycle based methods for sustainable product development. Int J Life Cycle Assess 8(3):157–159
Klöpffer W (2005) The critical review process according to ISO 14040–43. An analysis of the standards and experiences gained in their application. Int J Life Cycle Assess 10(2):98–102
Klöpffer W (2006) The role of SETAC in the development of LCA. Int J Life Cycle Assess 11(Special issue 1):116–122
Klöpffer W (2008) State-of-the-art in life cycle sustainability assessment (LCSA). Life cycle sustainability assessment of products (with comments by Helias A. Udo de Haes, p. 95). Int J Life Cycle Assess 13(2):89–95
Klöpffer W (2012) The critical review of life cycle assessment studies according to ISO 14040 and 14044—origin, purpose and practical performance. Int J Life Cycle Assess 17(9):1087–1093
Kotaji S, Schuurmans A, Edwards S (eds) (2003) Life-cycle assessment in building and construction—a state-of-the-art report. ISBN 1-880611-59-7. SETAC Press, Pensacola
O'Brien M, Doig A, Clift R (1996) 6th SETAC Europe meeting: LCA—selected papers. Social and environmental life cycle assessment (SELCA). Approach and methodological development. Int J Life Cycle Assess 1(4):231–237
Owens JW (1998) Life cycle impact assessment—the use of subjective judgments in classification and characterization. Int J Life Cycle Assess 3(1):43–46
Owens JW (1999) Why life cycle impact assessment is now described as an indicator system. Int J Life Cycle Assess 4(2):81–86
Poulsen PB, Jensen AA, Antonsson A-B, Bengtsson G, Karling M, Schmidt A, Brekke O, Becker J, Verschoor AH (eds) (2004) The working environment in LCA. ISBN 1-880611-68-6. SETAC Press, Pensacola
Rebitzer G, Ekvall T (eds) (2004) Scenarios in life-cycle assessment. ISBN 1-880611-57-0. SETAC Press, Pensacola
Todd JA, Curran MA (eds) (1999, Jul) Streamlined life-cycle assessment—a final report from the SETAC North America streamlined LCA workgroup. SETAC and SETAC foundation for environmental education
Töpfer K (2002) Editorial for Int J LCA on the launch of the UNEP/SETAC Life Cycle Initiative. Int J Life Cycle Assess 7(4):191
Udo de Haes HA (ed) (1996, Sept) Towards a methodology for life cycle impact assessment. ISBN 90-5607-005-3. Published by SETAC Europe, Brussels
Udo de Haes HA, Jensen AA, Klöpffer W, Lindfors L-G (eds) (1994) Integrating impact assessment into LCA proceedings of the LCA symposium held at the fourth SETAC Europe congress, 11–14 April 1994, Brussels. Published by Society of Environmental Toxicology and Chemistry-Europe, Brussels

Udo de Haes HA, Jolliet O, Finnveden G, Hauschild M, Krewitt W, Müller-Wenk R (1999a) Best available practice regarding impact categories and category indicators in life cycle impact assessment. Background document for the second working group on life cycle impact assessment of SETAC Europe (WIA-2, Part 1). Int J Life Cycle Assess 4(2):66–74

Udo de Haes HA, Jolliet O, Finnveden G, Hauschild M, Krewitt W, Müller-Wenk R (1999b) Best available practice regarding impact categories and category indicators in life cycle impact assessment. Background document for the second working group on life cycle impact assessment of SETAC Europe, Part 2. Int J Life Cycle Assess 4(3):167–174

Udo de Haes HA, Finnveden G, Goedkoop M, Hauschild M, Hertwich EG, Hofstetter P, Jolliet O, Klöpffer W, Krewitt W, Lindeijer E, Müller-Wenk R, Olsen SI, Pennington DW, Potting J, Steen B (eds) (2002a) Life-Cycle impact assessment: striving towards best practice. ISBN 1-880611-54-6. SETAC Press, Pensacola

Udo de Haes HA, Jolliet O, Norris G, Saur K (2002b) Life cycle initiative—background, aims and scope. Int J Life Cycle Assess 7(4):192–195

Valdivia S, Ugaya CML, Hildenbrand J, Traverso M, Mazijn B, Sonnemann G (2012) A UNEP/SETAC approach towards a life cycle sustainability assessment—our contribution to Rio+20. Int J Life Cycle Assess 18(9):1673–1685. doi:10.1007/s11367-012-0529-1

Chapter 3
The International Standards as the Constitution of Life Cycle Assessment: The ISO 14040 Series and its Offspring

Matthias Finkbeiner

Abstract The establishment of the international standards of Life Cycle Assessment—LCA (ISO 14040 series) led to worldwide acceptance of LCA. The ISO standards of LCA (ISO 14040 and ISO 14044) are the only globally relevant international standard documents on LCA which are broadly referenced by users and other standardization processes. Thus, they represent the constitution of LCA.

This chapter opens with an outline of the historical development of the international LCA standardization process and ends with an outlook on the future. The main part deals with the core standards and the spin-off standards of LCA. The core standards are ISO 14040—Environmental Management—Life Cycle Assessment—Principles and Framework and ISO 14044—Environmental Management—Life Cycle Assessment—Requirements and Guidelines.

Based on these classical LCA standards, 'new' approaches have recently been developed which have led to several spin-off-standards. They cover issues such as:

- 'Single-issue-LCAs' like carbon footprinting (ISO 14067) or water footprinting (ISO 14046),
- 'Beyond environment-LCAs' like life cycle costing, social LCA and eco-efficiency assessments (ISO 14045) or even life cycle sustainability assessments,
- 'Beyond product-LCAs' like Organizational LCAs (ISO 14072) or sector-based IO-LCAs and
- 'Beyond quantification-LCAs' like type III environmental product declarations (ISO 14025) or other types of environmental labels and claims.

Keywords Carbon footprint · Eco-efficiency · History of life cycle assessment standards · International life cycle assessment standards · ISO 14040 series · Organizational LCA · Spin-off-standards · Standards of life cycle assessment · Water footprint

---

M. Finkbeiner (✉)
Department of Environmental Technology, Chair of Sustainable Engineering,
Technische Universität Berlin, Office Z1, Strasse des 17. Juni 135,
10623, Berlin, Germany
e-mail: matthias.finkbeiner@tu-berlin.de

W. Klöpffer (ed.), *Background and Future Prospects in Life Cycle Assessment*,
LCA Compendium – The Complete World of Life Cycle Assessment,
DOI 10.1007/978-94-017-8697-3_3, © Springer Science+Business Media Dordrecht 2014

1 Introduction

Standards play an important role in business and everyday life. They represent a consensus on good practice and state of the art. This applies to all kinds of technical topics—including life cycle assessment (LCA). International standards for LCA were developed since the nineties as part of the ISO 14000 family of environmental management standards. In 2010, the drivers for this development were summarized in a brochure of the responsible ISO technical committee (ISO 2010) due to the observation"…that organizations around the world, as well as their stakeholders, are becoming increasingly aware of the need for environmental management, socially responsible behaviour and sustainable development. Accordingly, as the proactive management of environmental aspects converges with enterprise risk management, corporate governance, sound operational practices and financial performance, international standards are becoming increasingly important for organizations to work towards common and comparable environmental management practices to support the sustainability of their organizations, products, and services. It is the role of such standards to be technically credible, to fulfill stakeholder needs, to facilitate the development of uniform requirements, to promote efficiencies, to support compliance, to enhance investor confidence and to lead to continual improvement".

ISO is the International Organization for Standardization. It has a membership of over 160 national standards institutes from countries large and small, industrialized, developing and in transition, in all regions of the world. ISO's portfolio of more than 18,000 standards provides practical tools for all three dimensions of sustainable development: economic, environmental and societal. ISO technical committee ISO/TC 207 'Environmental management' is responsible for developing and maintaining the ISO 14000 family of standards. The committee's current portfolio consists of more than 20 published international standards and other types of normative documents, with about another ten new or revised documents in preparation. ISO/TC 207 was established in 1993, as a result of ISO's commitment to respond to the complex challenge of 'sustainable development' articulated at the 1992 United Nations Conference on Environment and Development in Rio de Janeiro. Membership of ISO/TC 207 is among the highest of any ISO technical committee and is both broad and diverse in representation, which are two key indicators of the worldwide interest in its work. National delegations of environmental experts from over 100 countries participate in ISO/TC 207, including over 25 developing countries (ISO 2010).

That committee within ISO/TC207 dealing with LCA is called Subcommittee 5 or in short: ISO/TC207/SC5. It was established right from the beginning of TC207. So far, the leadership of SC5 has always been provided by Germany: 1993–2000, Manfred Marsmann, 2000–2007, Hans-Jürgen Klüppel, 2007–now, Matthias Finkbeiner. The Secretariat has always been regulated by the national French standardization body AFNOR. To implement ISO´s 'twinning' policy, Reginald Tan from Singapore has been serving as co-chair of SC5 since several years.

Within this introductory chapter, the history of LCA standards development (Sect. 1.1), the relevance of ISO standards on LCA (Sect. 1.2) and the standardization process (Sect. 1.3) itself will be introduced. This will be followed by a high-level description of the core standards of LCA: ISO 14040 and ISO 14044 (Sect. 2) and the the spin-off standards from ISO 14040 and 14044 (Sect. 3). Finally, the future standards based on ISO 14040 and ISO 14044 (Sect. 4) are introduced before this article concludes with an outlook (Sect. 5).

1.1 History of LCA Standards Development

This section will describe the history of the development of the international standards of LCA in three periods: the early days (see Sect. 1.1.1), the first revision (see Sect. 1.1.2) and the proliferation (see Sect. 1.1.3).

1.1.1 The Early Days

The standardization process of the early days was a real challenge, because in many methodological issues there was no real consensus when it started. Despite some important references serving as seed documents, especially the so-called 'Code of Practice' (SETAC 1993) from SETAC (Society of Environmental Toxicology and Chemistry), particularly the methodologies of impact assessment and interpretation had to be standardized in parallel to the ongoing scientific development. At that time SETAC was the most relevant platform for LCA discussions and methodology development (see this volume, Chap. 2).

Initially, the standardization process within ISO/TC207/SC5 was organized in five separate working groups (WGs) (Marsmann 1997, 2000; Marsmann et al 1997). WG 1 on principles and guidelines established in 1997 the first and basic document of the emerging ISO 14040-series, i.e. ISO 14040—Environmental Management—Life Cycle Assessment—Principles and Guidelines (ISO 14040 1997). WG 2 dealt with generic and WG 3 with specific aspects of the life cycle inventory. In 1998, both groups together established ISO 14041—Environmental Management—Life Cycle Assessment—Goal and Scope Definition and Inventory Analysis (ISO 14041 1998). WG 4 was working on the assessment of environmental impacts and their valuation and produced in 2000 ISO 14042—Environmental Management—Life Cycle Assessment—Life Cycle Impact Assessment (ISO 14042 2000). WG 5 was dealing with the interpretation phase and managed to publish in 2000 ISO 14043—Environmental Management—Life Cycle Assessment—Life Cycle Interpretation (ISO 14043 2000).

The publication of these first international standards of LCA was an important milestone for the application of LCA and an essential step to consolidate procedures and methods. However, the complex structure of the working groups, the partly parallel, partly serial development of the documents and the long time needed for getting them published reveal the challenges to achieve international consensus. In addition, the parallel development of documents in different working groups has led to some inconsistencies between the first generation of standards that have been corrected in the first revision described in the following Sect. 1.1.2. However, despite such improvements, the key structure of the method, the four phases of LCA and the key requirements stood the test of time exceptionally well. The first revision reconfirmed, to a very large degree, the validity of the technical content of the first generation of standards. As a matter of fact, this clearly documents the outstanding work of the standardization pioneers in this first generation of LCA standards. They were ahead of the times in establishing these standards well before the years of the LCA boom. The relevance of this accomplishment gets even more obvious, if we take into account the standardization trials on carbon footprinting. Without proper LCA standards in place, the failure to deliver an international carbon footprint standard with sufficient speed to market (see Sect. 4.3) would be even more critical.

1.1.2 The First Revision

In the paper of Finkbeiner et al. (2006) about the new international standards for Life Cycle Assessment, ISO 14040 and ISO 14044 (Finkbeiner et al. 2006), the revision process and the main improvements achieved during the revision have been described in detail. This section is based on this paper and highlights some of the major changes made.

After the application experience of the first version of the standards, ISO/TC207/ SC5 started a consultation on the need and the strategy of a revision of the first generation of standards. A consensus was achieved on the following four key objectives:

- Increase readability by compiling only two documents/ merging different documents/ reorganising the current standards, but
  - Keep the technical content (only improvements are acceptable),
  - Keep the consensus/ balance,
  - Keep the requirements.
- Address applications of LCA (life cycle thinking; relation to ecolabels, design for environment (DfE), life cycle management, etc.).
- Inclusion of economic and social aspects are beyond the scope of TC207, but links should be addressed.
- Give guidance/ training for application in industry, government, etc., especially in developing countries.

Not all of these issues could be handled within an international standardization process. However, most of the issues could be solved by a revision of the standards. To explore this possibility and with a focus to improve the readability of the ISO 14040 series, a new ad-hoc group was created in June 2002 to review the ISO 14040/41/42/43 standards. The mandate of the ad-hoc group was to seek consensus on a possible way for a revision of these standards (boundaries of the revision, structure, contents, etc.). The mandate also demanded to explore if there is a consensus to develop the corresponding New Work Item Proposals (NWIPs) with accompanying working documents.

The ad-hoc group, consisting of 21 international experts and co-chaired by Atsushi Inaba and Matthias Finkbeiner, had one meeting and achieved a consensus on a possible way of revision of the standards; it also developed the necessary elements for the corresponding NWIPs which were presented to ISO/TC207/SC5 in July 2003.

The scope of the proposed work items was to begin immediately with the revision of the standards ISO 14040, 14041, 14042 and 14043, with the objective of improving readability, while leaving the requirements and technical content unaffected, except for errors and inconsistencies. It was the intention:

1. to gather all requirements ('shalls') in one new standard, keeping the structure of 'goal and scope', 'inventory', 'impact assessment' and 'interpretation' as separate chapters,
2. to maintain ISO 14040 as a framework document, but transferring all requirements ('shalls') to the new standard, adding to ISO 14040 a requirement ('shall') of compliance with the requirements ('shalls') of the new standard.

This proposal was justified with regard to applicability and readability due to the request of several member bodies for improvement, because the existing documents were partly not consistent, partly not clear or even ambiguous. In addition to language improvement, a merging of standards was requested by some member bodies to make them more readable.

As indicated in the scope of the NWIP, it was proposed, to fulfill this need by two new standards: a revised ISO 14040 standard ('Environmental Management—Life Cycle Assessment—Principles and Framework') (ISO 14040 2006) and a new standard 14044 containing all requirements ('Environmental Management—Life Cycle Assessment—Requirements and Guidelines') (ISO 14044 2006).

The voting of the international member bodies on this proposal in the autumn of 2003 revealed an unanimous result (no negative vote, two abstentions). Therefore, a new working group WG6 (with more than 50 international experts, co-chaired by Atsushi Inaba (Japan), Reginald Tan (Singapore) and Matthias Finkbeiner (Germany), Secretariat provided by Kim Christiansen (Denmark)) was created to accomplish the revision of the standards according to the scope of the NWIPs. WG6 was working very efficiently and in good team spirit. Despite the fact that it had to deal with in total 1,900 comments, the work was accomplished with basically complete consensus in the minimum number of WG meetings and a few months ahead of schedule. Even though the scope for the revision was rather restrictive, several changes were made compared to the first generation of standards.

An obvious formal change due to the revision is the reduced number of standards, the reduced number of annexes and the reduced number of pages that contain requirements. All these changes were intended to increase the readability and accessibility of the standards. For the practitioners of LCA, this means that the technical requirements can be found in one document (instead of previously four) and that they are condensed on 26 pages (instead of 44 previously).

Next to the more formal changes, some technical modifications were made as well. Generally, the main technical content of the previous standards was reconfirmed to be still valid. Many important issues of fundamental importance, e.g. allocation, requirements for comparative assertions or the phases of LCA were not changed. This was both not the intention of the revision and not found to be justified during the revision process. However, still some technical changes were made. The modified technical content is in line with the previous requirements and serves mainly as a clarification of the technical content, and as a correction of errors and inconsistencies. It includes, e.g., the addition of several definitions (e.g. product, process, etc.), the addition of principles for LCA, clarifications concerning

- LCA intended to be used in comparative assertions intended to be disclosed to the public,
- system boundaries,
- the critical review panel, and
- the addition of an annex about applications.

As an example, both the previous and the new ISO 14040 have in the title 'principles and framework', but the previous version did not include any principles. To remove this inconsistency, the following principles were added to the new ISO 14040:

- Life cycle perspective.
- Environmental focus.
- Relative approach and functional unit.
- Iterative approach.
- Transparency.
- Comprehensiveness.
- Priority of scientific approach.

It is explained that these principles are fundamental and should be used as guidance for decisions relating to both the planning and the conducting of an LCA.

The revised standards were approved by unanimous vote which means that they represent a complete consensus of all countries and stakeholders. The versions of ISO 14040 and 14044 developed in 2006 are still valid today. As part of the systematic review procedure of ISO standards, there was an inquiry on the need for revision to all member bodies in 2009. The result of the inquiry was an almost unanimous confirmation of the existing standards.

While the standards are sometimes criticized by some stakeholders (especially from academia) for not being specific enough on certain issues, they do represent the global consensus on those methodological features for which such a consensus exists. More specific stipulations on, e.g., allocation procedures or a default set of impact categories, let alone a particular impact assessment method might be desired by some stakeholders, but there is no global stakeholder consensus on that. It makes no sense to blame the standards for this, as it is the natural result of the very democratic procedure to develop an ISO standard. The fact, that ISO 14040 and 14044 represent such a strong consensus among both private and public users of LCA and that they are the only globally relevant international standards of LCA, makes them so relevant. This was particularly demonstrated by their important role in the proliferation of standards based on LCA, which is addressed in the next Sect. 1.1.3.

1.1.3 The Proliferation

Soon after the publication of the revised standards, LCA started to boom. While part of this growth came from increased application and implementation of LCA itself in both private and public decision-making, an additional momentum was generated by the development of 'new' approaches built on the basis of classical LCA:

- 'Single-issue-LCAs' like carbon footprinting or water footprinting,
- 'beyond environment-LCAs' like life cycle costing, social LCA and eco-efficiency assessments or even life cycle sustainability assessments,
- 'beyond product-LCAs' like scope 3 type LCAs of organizations or sector-based IO-LCAs and
- 'beyond quantification-LCAs' like type III environmental product declarations or other types of environmental labels and claims.

While some of these additional standards are part of the ISO/TC207 family, additional public and private standardization bodies tried to penetrate the market with their products. Especially, the carbon footprint discussions led to a huge proliferation of different guidelines and standards. In the editorial 'Carbon footprinting—opportunities and threats' (Finkbeiner et al. 2006) which was published to announce a particular carbon footprint section in the International Journal of Life Cycle Assessment, the following non-exclusive list of initiatives was given:

- ISO 14067 on Carbon Footprint of Products.
- The World Business Council for Sustainable Development (WBCSD) and the World Resources Institute (WRI) developed two standards under their Greenhouse Gas Protocol Product/Supply Chain Initiative: A Product Life Cycle Accounting and Reporting Standard and a Corporate Accounting and Reporting Standard: Guidelines for Value Chain (Scope 3) Accounting and Reporting.
- The UNEP/SETAC Life Cycle Initiative launched a project group on carbon footprinting.
- The British Standards Institution published a Publicly Available Specification (PAS) to specify requirements for assessing the life cycle greenhouse gas emissions (GHG) of goods and services. The development of this PAS was co-sponsored by the Carbon Trust and the Department for Environment, Food and Rural Affairs (PAS 2050 2011).

- The Japanese Ministry of Economy, Trade and Industry (METI) launched a carbon footprint trial project, and a Technical Specification 'General principles for the assessment and labelling of Carbon Footprint of Products' was issued.
- Many more initiatives were launched, in Korea, the European Union, France, Germany, New Zealand, etc.

In Sect. 4, the future standards within the ISO 14000 series are introduced which are built on the LCA standards.

1.2 Relevance of ISO Standards on LCA

In the early days of LCA, the results of the studies were often apparently biased by vested interests of the study commissioners. While the general idea and concept of LCA was appealing to many stakeholders right from the start, the credibility of the method was severely damaged by such misuse. These 'wild-west' times of LCA had been overcome when the international standards of LCA were published to improve the quality of LCAs and to hinder wrong claims about the environmental superiority of products.

Before the establishment of the ISO standards of LCA, governments were reluctant to apply LCA for their policy development due to a lack of commonly accepted procedures and methods. Companies often had a risk-aversive strategy towards LCA because they were either afraid of market distortions of unjustified claims by competitors or barriers of trade, or of mandatory reporting requirements by public policy. These fears were partly amplified by the tendency of some LCA practitioners to oversell the tool. For some of them LCA was not any longer a tool, it was more a religion to determine what is good and what is evil. During that period in the nineties, LCA practitioners from academia and consultancy were typically belonging to different schools fighting about the right way to do LCA, the right impact assessment approach, the best LCA software, and so on.

The international standardization of LCA achieved a much clearer perspective and a much more sober view what LCA can do, but—at least likewise important—also what it can*not* do. It established a common language of terms and key methodological requirements, but it did not fix 'a one size fits all-LCA'. By giving the users of LCA an equally important voice as the providers of LCA, the consensus achieved in the standards did not make everybody happy, but it reached a fairly strong global consensus on the basic rules and framework of LCA. In addition, the standards made the limitations of LCA transparent and provided fairly strict requirements for the most contentious application of LCA, the so-called comparative assertions intended to be disclosed to the public. As such, the establishment of the international standards was of utmost importance for the broad acceptance of LCA all around the world and by all stakeholders. The ISO standards of LCA are until today the one and only globally relevant international standard documents which are broadly referenced by users and other standardization processes.

1.3 ISO's Standardization Process

According to ISO, "an ISO Standard is a normative document, developed according to consensus procedures, which has been approved by the ISO membership and P-members of the responsible committee in accordance with the ISO/IEC Directives. ISO standards are developed by groups of experts, within technical committees (TCs). TCs are made up of representatives of industry, NGOs, governments and other stakeholders, who are put forward by ISO's members. Each TC deals with a different subject, for example there are TCs focusing on screw threads, shipping technology, food products and many, many more.

ISO's full members (member bodies, i.e. national standardization organizations) can decide if they would like to be a participating member (P-member) of a particular TC or an observing member (O-member). P-members participate actively in the work and have an obligation to vote on all questions submitted to vote within the technical committee. O-members follow the work as an observer but cannot make any comments about the development process or vote.

An ISO standard is developed by a panel of experts, within a technical committee. Once the need for a standard has been established, these experts meet in a working group established for this purpose to discuss and negotiate a draft standard. As soon as a draft has been developed, it is shared with ISO members who are asked to comment and vote on it. If a consensus is reached, the draft becomes an ISO standard, if not it goes back to the technical committee for further edits" (ISO 2012a).

According to ISO, the standardization process is built on four key principles (ISO 2012b):

- **"ISO standards respond to a need in the market**
  ISO does not decide when to develop a new standard. Instead, ISO responds to a request from industry or other stakeholders such as consumer groups. Typically, an industry sector or group communicates the need for a standard to its national member who then contacts ISO.
- **ISO standards are based on global expert opinion.**
  ISO standards are developed by groups of experts from all over the world that are part of larger groups called technical committees. These experts negotiate all aspects of the standard, including its scope, key definitions and content.
- **ISO standards are developed through a multi-stakeholder process**
  The technical committees are made up of experts from the relevant industry, but also from consumer associations, academia, NGOs and government.
- **ISO standards are based on a consensus**
  Developing ISO standards is a consensus-based approach and comments from stakeholders are taken into account."

2 The Core Standards of LCA: ISO 14040 and ISO 14044

As described in Sect. 1.1.2, ISO 14040 2006 and ISO 14044 2006 are the core standards of LCA that are still valid today. The current ISO 14040 is a framework and guidance standard, while ISO 14044 contains all technical requirements and guidelines on these. Therefore, ISO 14040 provides a more general, introductory reading of the concept and outline of LCA including its principles. For the LCA practitioner, ISO 14044 is the operational document including all requirements for ISO compliant LCA studies. This section is by no means able to replace reading the actual standard documents. It is rather intended to give a flavor of the key features of the standards and supposed to proselytize those who so far resisted the core standards of LCA.

According to Sect. 4.3 of ISO 14040 (ISO 14040 2006), the following aspects are defined as key features of the LCA methodology:

- "LCA assesses, in a systematic way, the environmental aspects and impacts of product systems, from raw material acquisition to final disposal, in accordance with the stated goal and scope;
- The relative nature of LCA is due to the functional unit feature of the methodology;
- The depth of detail and time frame of an LCA may vary to a large extent, depending on the goal and scope definition;
- Provisions are made, depending on the intended application of the LCA, to respect confidentiality and proprietary matters;
- LCA methodology is open to the inclusion of new scientific findings and improvements in the state-of-the-art of the technique;
- Specific requirements are applied to LCA that are intended to be used in comparative assertions intended to be disclosed to the public;
- There is no single method for conducting LCA. Organizations have the flexibility to implement LCA,..., in accordance with the intended application and the requirements of the organization;
- LCA is different from many other techniques (such as environmental performance evaluation, environmental impact assessment and risk assessment) as it is a relative approach based on a functional unit; LCA may, however, use information gathered by these other techniques;
- LCA addresses potential environmental impacts; LCA does not predict absolute or precise environmental impacts due to
  - the relative expression of potential environmental impacts to a reference unit,
  - the integration of environmental data over space and time,
  - the inherent uncertainty in modelling of environmental impacts, and
  - the fact that some possible environmental impacts are clearly future impacts;
- The LCIA phase, in conjunction with other LCA phases, provides a system-wide perspective of environmental and resource issues for one or more product system(s);

- LCIA assigns LCI results to impact categories; for each impact category, a life cycle impact category indicator is selected and the category indicator result (indicator result) is calculated; the collection of indicator results (LCIA results) or the LCIA profile provides information on the environmental issues associated with the inputs and outputs of the product system;
- There is no scientific basis for reducing LCA results to a single overall score or number, since weighting requires value choices;
- Life cycle interpretation uses a systematic procedure to identify, qualify, check, evaluate and present the conclusions based on the findings of an LCA, in order to meet the requirements of the application as described in the goal and scope of the study;
- Life cycle interpretation uses an iterative procedure both within the interpretation phase and with the other phases of an LCA;
- Life cycle interpretation makes provisions for links between LCA and other techniques for environmental management by emphasizing the strengths and limits of an LCA in relation to its goal and scope definition."

These key features describe the main aspects of LCA according to the ISO-standards. A particular feature mentioned there are the additional requirements for LCAs that are intended to support comparative assertions intended to be disclosed to the public. This application has potentially strong implications on third parties. As a consequence, ISO 14044 provides a set of particular requirements for these types of studies:

- The equivalence of the systems being compared shall be evaluated before interpreting the results. Systems shall be compared using the same functional unit and equivalent methodological considerations such as performance, system boundary, data quality, allocation procedures, decision rules on evaluating inputs, and outputs and impact assessment. Any differences between systems regarding these parameters shall be identified and reported.
- While an LCI study without impact assessment is a feasible choice for any other application, an LCIA is required for comparisons intended to be used in comparative assertions to be disclosed to the public.
- The LCIA shall employ a sufficiently comprehensive set of category indicators. The comparison shall be conducted by category indicator.
- Weighting shall not be used in LCA studies intended to be used in comparative assertions intended to be disclosed to the public.
- Several data quality requirements and sensitivity analyses are required and not only recommended.
- In order to decrease the likelihood of misunderstandings or negative effects on external interested parties, a critical review of a panel of interested parties is mandatory, whereas critical reviews are just recommended for all the other applications.
- Finally, specific reporting requirements apply as described in paragraph 5.3 of ISO 14044.

As any reputable LCA practitioner is supposed to get acquainted with the core standards of LCA rather sooner than later, we need not go into further details including all principles, requirements, guidelines, annexes on applications, examples of data collection sheets and examples of interpretation. However, the lesser known spin-off standards (see Sect. 3) of ISO 14040 and ISO 14044 as well as the currently developed future standards (see Sect. 4) justify a brief introduction.

3 The Spin-off Standards

The core standards of LCA quickly generated offspring, i.e. standards resulting from application of or as additional guidance to the ISO 14040 series of standards. One spin-off standard was developed outside ISO TC 207/SC5, because it is part of the ISO 14020 series of ecolabelling standards. However, as ISO 14025 on type III environmental declarations (ISO 14025 2006) provides basically a standardized reporting format for LCAs; it is briefly described in Sect. 3.1. Already in parallel to the development of the first generation of core standards of LCA, i.e. ISO 14040–43, the discussion started to supplement these requirement standards with non-normative documents (Technical Reports) that provide examples for their application. The resulting documents ISO/TR 14047 (ISO/TR 14047 2012) and ISO/TR 14049 (ISO/TR 14049 2012) are introduced in Sects. 3.2 and 3.4, respectively. The third spin-off document ISO/TS 14048 (ISO/TS 14048 2002), which is described in Sect. 3.3 deals with the issue of data documentation format.

3.1 ISO 14025—Type III Environmental Product Declarations

The ISO 14020 series differentiates between three types of environmental labels and declarations. Type I labels are the classical ecolabels like the German Blue Angel for providing a clear indication of environmental superiority to the consumer. Type II labels and claims are rather flexible and particularly focus on self-declared claims without third party verification. ISO type III environmental declarations provide quantified environmental data using predetermined parameters and, where relevant, additional environmental information. Most importantly, the predetermined parameters are based on the ISO 14040 series of standards, i.e. LCA. In a nutshell, such environmental product declarations (EPDs) are small environmental reports of a product reporting its LCA.

The latter is covered by the standard ISO 14025—Environmental Labels and Declarations—Type III Environmental Declarations—Principles and Procedures (ISO 14025 2006). This standards was published 2006 and was built on a Technical Report ISO/TR 14025 which was first issued in 2000. This standard establishes the principles and specifies the procedures for developing Type III environmental declaration programs and Type III environmental declarations. It specifically establishes the use of the ISO 14040 series of standards in the development of Type III environmental declaration programs and Type III environmental declarations.

Type III environmental declarations are primarily intended for use in business-to-business communication, but their use in business-to-consumer communication is not precluded.

On a technical level, ISO 14025 developed the concept of the so-called product category rules (PCRs). PCRs represent basically a predetermined goal and scope definition for a particular product group and are intended to achieve comparability within a set of products. The concept of PCRs gained significant importance in the current discussions on carbon footprint labels and are nowadays recognized as a relevant and feasible option to further specify the generic, cross-sectorial requirements of the LCA or footprinting standards for particular product groups.

3.2 ISO 14047—Examples of Impact Assesement

This document is officially called 'ISO/TR 14047: 2012 Environmental Management—Life Cycle Assessment—Illustrative examples on how to apply ISO 14044 to impact assessment' (ISO/TR 14047 2012). The current version is an editorially improved version of the original document from 2003.

"The purpose of this Technical Report is to provide examples to illustrate current practice of life cycle impact assessment according to ISO 14044 2006. These examples are only a sample of all possible examples that could satisfy the provisions of the standard. They offer 'a way' or 'ways' rather than the 'unique way' of applying the ISO 14044 2006. They reflect the key elements of the life cycle impact assessment (LCIA) phase of the LCA. The examples presented in this TR are not exclusive and other examples exist to illustrate the methodological issues described" (ISO/TR 14047 2012).

As there was no technical update during the revision, several of the examples given do not necessarily represent the latest state-of-the-art in impact assessment. In general, the document was definitely beneficial for some users, but has probably not really achieved a strong impact on the LCA community.

3.3 ISO 14048—Data Documentation Format

This Technical Specification 'Environmental Management—Life Cycle Assessment—Data Documentation Format' (ISO/TS 14048 2002) provides the requirements and a structure for a data documentation format to be used for transparent and unambiguous documentation and exchange of Life Cycle Assessment (LCA) and Life Cycle Inventory (LCI) data, thus permitting consistent documentation of data, reporting of data collection, data calculation and data quality, by specifying and structuring relevant information.

"The data documentation format specifies requirements on division of data documentation into data fields, each with an explanatory description. The description of each data field is further specified by the structure of the data documentation format.

The document intends to support LCA use and development, and is aimed primarily for data suppliers, LCA practitioners and LCA information system developers. The data documentation format is also intended to facilitate the exchange of LCI data without loss of transparency, even though the specification does not provide specific requirements for implementation of data exchange. The specification, explanation and implementation of the data documentation format are described in different parts of the document as follows:

- Clause 5 covers the specification and structure of the data documentation format and the names of all of the data fields;
- Clause 6 covers the specification of the data types used in the data documentation format;
- Clause 7 covers the specification of nomenclatures used in the data documentation format;
- Annex A contains formatting requirements and explanatory descriptions of each data field to help the user understand which information to place in each data field;
- Annex B contains a detailed example of the use of the data documentation format" (ISO/TS 14048 2002).

Due to the technical nature of the document, the relevance for the average LCA practitioner and user is somewhat limited. However, for database providers and software developers, ISO/TS 14048 serves as a useful reference.

3.4 ISO 14049—Examples of Inventory Analysis

The full title of this document is 'ISO/TR 14049: 2012 Environmental Management—Life Cycle Assessment—Illustrative examples on how to apply ISO 14044 to goal and scope definition and inventory analysis' (ISO/TR 14049 2012).

"This Technical Report provides examples about practices in carrying out a Life Cycle Inventory Analysis (LCI) as a means of satisfying certain provisions of ISO 14044. These examples are only a sample of the possible cases satisfying the provisions of ISO 14044. They offer 'a way' or 'ways' rather than the 'unique way' for the application of ISO 14044. These examples reflect only portions of a complete LCI study.

- Apart from some general content, the TR focuses on
- Examples of developing functions, functional units and reference flows
- Examples of distinguishing functions of comparative systems
- Examples of establishing inputs and outputs of unit processes and system boundaries

- Examples of avoiding allocation
- Examples of allocation
- Example of applying allocation procedures for recycling
- Examples of conducting data quality assessment
- Examples of performing sensitivity analysis (ISO/TR 14049 2012)."

Compared to ISO/TR 14047, the examples presented here stood the test of time fairly well and the document is still quite relevant today. From the author's non-representative experience, ISO/TR 14049 is the most popular and most used of the three spin-off-standards described in this section.

4 The Future Standards Based on ISO 14040/44

In this section, the future standards based on ISO 14040/44 are introduced. First, the just published ISO 14045 on eco-efficiency assessment is described in Sect. 4.1. This standard goes beyond the purely environmental perspective of LCA and adds the economic perspective into the assessment. In contrast to such broadening of the scope of LCA, the single-issue or footprinting standards have become popular recently. The upcoming ISO 14046 on water footprint is addressed in Sect. 4.2 while carbon footprinting according to ISO/TS 14067 is covered in Sect. 4.3. The final two documents ISO 14071 on critical review (Sect. 4.4) and ISO 14072 on the use of LCA for organizations (Sect. 4.5) provide additional specifications for one element of LCA, respectively guidance for the use of LCA not only on the product level, but also organization level. Because all numbers of the ISO 14040 series were already used, the two latter documents are developed within the new, additional number set for LCA, i.e. the ISO 14070 series (Finkbeiner 2013).

4.1 ISO 14045—Eco-Efficiency Assessment

The standard 'ISO 14045: 2012 Environmental Management—Eco Efficiency Assessment of Product Systems—Principles, Requirements and Guidelines' represents an important step due to a broader focus beyond environmental issues only (ISO 14045 2012). "Eco-efficiency assessment is a quantitative management tool which enables the consideration of life cycle environmental impacts of a product system alongside its product system value to a stakeholder.

Within eco-efficiency assessment, environmental impacts are evaluated using Life Cycle Assessment (LCA) as prescribed by other International Standards (ISO 14040, 14044). Consequently, eco-efficiency assessment shares with LCA many important principles such as life cycle perspective, comprehensiveness, functional unit approach, iterative nature, transparency and priority of scientific approach.

The value of the product system may be chosen to reflect, for example, its resource, production, delivery or use efficiency, or a combination of these. The value may be expressed in monetary terms or other value aspects.

The key objectives of this International Standard are to:

- establish clear terminology and a common methodological framework for eco-efficiency assessment;
- enable the practical use of eco-efficiency assessment for a wide range of product (including service) systems;
- provide clear guidance on the interpretation of eco-efficiency assessment results;
- encourage the transparent, accurate and informative reporting of eco-efficiency assessment results" (ISO 14045 2012).

4.2 *ISO 14046—Water Footprint*

As mentioned by Berger and Finkbeiner (2012), water footprinting is now a priority in current sustainability discussions after having been neglected for many years due to a lack of both awareness and appropriate methods for accounting and assessing water use and consumption. There is currently not the one and only water footprint method but different approaches to analyze the water use and consumption of organizations or along product life cycles. Next to stand-alone methods, such as virtual water, the method of the Water Footprint Network, the global water tool, or the corporate water gauge, many methods were developed in an LCA context (Berger and Finkbeiner 2012). A review of these methods is provided by Berger and Finkbeiner (2010), even though it is already slightly outdated due to the dynamic developments in the field.

Both the increasing relevance of water footprinting and the diverse methods were the drivers to work on an international standard. The market need for such a standard is confirmed by the large participation in the working group dealing with it. The working group includes more than 100 experts from a diverse mix of countries from the developing and developed world, and both countries which are lucky to have a lot of water resources and countries that suffer from water scarcity. Due to the state of the art in water footprinting, it is premature to expect a standard that will fix THE method to do it. The first version of the standard is about agreeing on the relevant terminology and some key methodological issues and concepts. One of these issues is the discussion of volumetric versus impact-oriented water footprint methods. While the method of the Water Footprint Network, which deserves credit for bringing the issue on the agenda, uses an inventory of water volumes, it is nowadays broadly accepted that this is scientifically not sufficient to address the issue of water scarcity. This is acknowledged by ISO 14046 which currently defines a water footprint clearly on the impact level as "parameter(s) that quantify(ies) the potential environmental impacts related to water" (ISO 14046.CD.1 2012)).

According to the current committee draft document (ISO 14046.CD.1 2012), the scope is defined as "specifying principles, requirements and guidelines to assess and report the water footprints of products, processes and organizations based on life cycle assessment (LCA). The standard provides requirements and guidance for calculating and reporting a water footprint as a stand-alone assessment or as part of a more comprehensive environmental assessment. The water footprint is calculated as one impact indicator result or multiple impact indicator results."

Due to the large participation in the work and its relevance, the prediction of the publication date involves uncertainties. However, based on the current project plan, the publication of the document is expected for late 2014.

4.3 ISO/TS 14067—Carbon Footprint

As mentioned in Sect. 1.1.3, the topic of carbon footprinting contributed significantly to the growing use of life cycle based assessment tools on the one hand and to a proliferation of guides and 'standards' on the other. On the ISO level, the work on this topic is done by ISO/TC207/SC7 on Greenhouse Gas Management and has led to ISO/TS 14067 'Carbon footprint of products—Requirements and Guidelines for Quantification and Communication' (ISO/TS 14067 2013). According to the introduction of the document, "this International Technical Specification is based on existing ISO standards, e.g. ISO 14020, ISO 14025, ISO 14040 and ISO 14044 and aims to set more specific requirements for the quantification and communication of carbon footprints of products (CFP). Specific requirements apply where the CFP information is intended to be publicly available. This document is expected to benefit organizations, governments, communities and other interested parties by providing clarity and consistency for quantifying, communicating and verifying CFPs. Specifically, using life cycle assessment according to this International Standard with climate change as the single impact category may offer benefits through:

- providing requirements for the methods to be adopted in assessing the CFP;
- facilitating the tracking of performance in reducing GHG emissions;
- assisting in the creation of efficient and consistent procedures to provide CFP information to interested parties;
- providing a better understanding of the CFP such that opportunities for GHG reductions may be identified;
- providing CFP information to encourage changes in consumer behaviour which could contribute to reductions in GHG emissions through improved purchasing, use and disposal decisions;
- providing correct and consistent communication of CFPs which supports comparability of products in a free and open market;
- enhancing the credibility, consistency and transparency of the quantification, reporting and communication of the CFP;
- facilitating the evaluation of alternative product design and sourcing options, production and manufacturing methods, raw material choices, recycling and other end-of-life stages;

- facilitating the development and implementation of GHG management strategies and plans across product life cycles as well as the detection of additional efficiencies in the supply chain."

While the specification has grown to a lengthy document of over 50 pages, most of its content is just a repetition of content of previous standards. For the quantification part, a lot of content of ISO 14044 is copied into ISO/TS 14067. As a matter of fact, the additional CFP specific requirements for quantification are rather few and would easily fit on a handful of pages. While such a 'delta'-standard would have been more efficient, the working group wanted to develop a 'stand-alone' document. This was—amongst others (e.g. merging of the originally separate quantification and communication parts, lack of team spirit and knowledge on underlying standards, lack of process quality and leadership)—one of the reasons why the overall standardization process took much longer than it was supposed to be.

Originally, the publication of the standard was due in March 2011. The final publication of the document in 2013 was on the level of a Technical Specification and not on the level of an International standard, because the different committee drafts and draft international standards have been several times rejected in the voting of the national standardization bodies.

4.4 ISO 14071—Critical Review

The proposal to develop a Technical Specification ISO 14071 'Environmental Management—Life Cycle Assessment—Requirements and Guidelines for Critical Review Processes and Reviewer Competencies' was motivated from the discussion on conformity assessment of e.g. the carbon footprint standards or upcoming labelling initiatives. As part of these processes, different interested parties proposed different conformity assessment schemes including critical review according to ISO 14040 and ISO 14044, verification according to ISO 14025, but also the bureaucratic accountant approach according to greenhouse gas verification. The critical review approach was very successful within the LCA community. Despite the concise content in the standards, a common practice emerged in the market place that satisfied all stakeholders. For the mandatory case of comparative assertions disclosed to the public, but also in many cases for which a critical review is not mandatory, study commissioners decided to perform a critical review to improve their studies and to support credibility. One of the key success factors is that the system does not operate an accreditation scheme which tries to ensure quality by bureaucracy and in which verification bodies that can afford the overhead cost then send some inexperienced individuals actually doing the job. The critical review system ensured quality by making the individual reviewer accountable for the work and spending the resources on content rather than paper work.

While, this is generally accepted in the LCA world, the critical review system had a challenge to be argued 'against' the bloated verification documents of other schemes (sometimes even three—one for the verification process, one for verification bodies, one for the competence of verifiers). In order to document the well established critical review practice in a more formal way, the intention of this coming international technical specification is to provide requirements and guidelines for conducting a critical review and the competencies required. It will describe:

- details of a Critical Review process including clarification with regard to ISO 14044;
- guidelines to deliver the required Critical Review process linked to the goal of the LCA and its intended use;
- content and deliverables of the Critical Review process;
- guidelines to improve the consistency, transparency, efficiency and credibility of the Critical Review process;
- the required competencies for the reviewer(s) (internal, external and panel member);
- the required competencies to be represented by the panel as a whole.

The target is to provide a crisp and lean specification that documents the established best practice for performing critical reviews. The publication of the document is expected in 2014.

4.5 ISO 14072—Organizational LCA (OLCA)

The benefits and the potential of the life cycle approach are not limited to an application on products. While the LCA methodology was originally developed for products, its application on the organizational level is getting more and more relevant (Finkbeiner and König 2013). The discussions on carbon footprinting of companies including their upstream and downstream supply chains (the so-called 'scope 3' according to the GHG-Protocol) (Finkbeiner 2009) revealed that these 'life cycle' emissions can contribute significantly to the organizational footprint. The currently applied assessments mostly concentrate on a single aspect like carbon or water footprints. The purpose of this new standard is to present a general and comprehensive approach by adapting LCA methodology on organizations.

The document ISO 14072 is supposed to be a Technical Specification (TS) called Environmental management—Life cycle assessment—Requirements and guidelines for Organizational Life Cycle Assessment. The main goal is to provide additional guidance to organizations for an easier and more effective application of ISO 14040 and ISO 14044 on the organizational level including the advantages that LCA may bring to organizations, the system boundaries and the limitations regarding reporting, environmental declarations and comparative assertions. It is intended for any organization that has interest in applying LCA. It is not intended for ISO 14001 interpretation and covers the goals of ISO 14040 and 14044. The publication of the document is expected in 2014.

5 Summary and Outlook

The establishment of the international standards of LCA (ISO 14040 series) was crucial for the broad acceptance of LCA all around the world and by all stakeholders. The ISO standards of LCA (ISO 14040 and ISO 14044) are until today the one and only globally relevant international standard documents on LCA which are broadly referenced by users and other standardization processes. The standards contributed significantly to the transition of LCA from an academic toy or misused greenwashing machine towards a serious, robust and professional tool to support decision-making in public and private organizations.

They represent the constitution of LCA and should therefore be respected and protected by everyone. It is fair to ask for more specific stipulations in future versions, if global consensus evolves on such issues. If such a consensus does not exist, we have to be aware that asking for more sometimes leads to getting less than we already have—half a loaf is better than no bread.

Some future activities have already been highlighted in Sect. 4. They represent fairly well the future direction that the author anticipates at this point in time. We will have some additional standards that specify particular parts of LCA methodology (e.g. critical review, Organizational LCA), we will have some further standards on simplified LCA versions (e.g. carbon or water footprinting) and we will expand the environmental focus towards all three sustainability dimensions (resource efficiency, life cycle costing, social LCA, life cycle sustainability assessment). All these developments shall support the credible and robust use of LCA for real world decision-making in the sense of life cycle management and life cycle sustainability management (Finkbeiner 2011). "If we want to make sustainability happen as concrete reality in both public policy making and corporate strategies, sustainability cannot please everybody. This requires that we are able to address the question, how sustainability performance can be measured, especially for companies, products and processes. We have to be smart enough to be able to measure it or the real and substantial implementation of the sustainability concept will remain just wishful thinking. In order to achieve reliable and robust sustainability assessment results it is inevitable that the principles of comprehensiveness and life cycle perspective are applied" (Finkbeiner 2011). In systems that support participation of citizens and democracy, this requires commonly accepted rules. ISO 14040 and ISO 14044 achieved to be that for LCA in the last decade. The author sincerely hopes and expects that there are more global citizens and good reasons out there that ensure keeping such a basic law of LCA. It is by no means a guarantee for sustainable development, but it makes it more probable.

Appendix-Glossary

Core standards of LCA	ISO 14040 2006 and ISO 14044 2006
International standards of LCA	ISO 14040 series
ISO	International Organization for Standardization
ISO 14040	Environmental Management—Life Cycle Assessment—Principles and Framework (1997 and 2006)
ISO 14041	Environmental Management—Life Cycle Assessment—Goal and Scope Definition and Inventory Analysis (1998)
ISO 14042	Environmental Management—Life Cycle Assessment—Life Cycle Impact Assessment (2000)
ISO 14043	Environmental Management—Life Cycle Assessment—Life Cycle Interpretation (2000)
ISO 14044	Environmental Management—Life Cycle Assessment—Requirements and Guidelines (2006)
ISO/DIS	Draft International Standard
ISO/TC207/SC5	ISO/Technical Committee 207 (Environmental management)/ Subcommittee 5 (LCA)
ISO/TR	Technical report
ISO/TS	Technical specification
NWIPs	New Work Item Proposals
Spin-off standards	ISO 14025—Environmental Labels and Declarations, Type III: Environmental Declarations—Principles and Procedures (2006)
	ISO 14045—Environmental Management—Eco-efficiency Assessment of Product Systems—Principles, Requirements and Guidelines (2012)
	ISO/CD.1 14046—Water Footprint—Requirements and Guidelines (publication is expected in late 2014)
	ISO/TR 14047—Environmental Management—Life Cycle Assessment—Illustrative examples on how to apply ISO 14044 to impact assessment situations (2012)
	ISO/TS 14048—Environmental Management—Life Cycle Assessment—Data Documentation Format (2002)
	ISO/TR 14049—Environmental Management—Life Cycle Assessment—Illustrative examples on how to apply ISO 14044 to goal and scope definition and inventory analysis (2012)
	ISO/TS 14067—Carbon Footprint of Products—Requirements and Guidelines for Quantification and Communication (2013)
	ISO/TS 14071—Environmental Management—Life Cycle Assessment—Requirements and Guidelines for Critical Review Processes and Reviewer Competencies (publication is expected in 2014)
	ISO 14072 is supposed to be a Technical Specification called Environmental management—Life cycle assessment—Requirements and guidelines for Organizational Life Cycle Assessment (publication is expected in 2014)
WBCSD	World Business Council for Sustainable Development (WBCSD)

References

Berger M, Finkbeiner M (2010) Water footprinting—how to address water use in life cycle assessment? Sustainability 2(4):919–944
Berger M, Finkbeiner M (2012) Methodological challenges in volumetric and impact-oriented water footprints. J Ind Ecol 17(1):79–89. doi: 10.1111/j.1530-9290.2012.00495.x
Finkbeiner M (2009) Carbon footprinting—opportunities and threats. Int J Life Cycle Assess 14(2):91–94
Finkbeiner M (2011) Towards life cycle sustainability management. Springer, Berlin, (ISBN: 978-94-007-1898-2)
Finkbeiner M (2013) From the 40s to the 70s—the future of LCA in the ISO 14000 family. Int J Life Cycle Assess (2013) 18:1–4
Finkbeiner M, Inaba A, Tan RBH, Christiansen K, Klüppel H-J (2006) The new international standards for life cycle assessment: ISO 14040 and ISO 14044. Int J Life Cycle Assess 11(2):80–85
Finkbeiner M, König P (2013) Carbon Footprint and Life Cycle Assessment of Organizations. Journal of Environmental Accounting and Management 1(1):55–63
ISO (2010) Environmental management—the ISO 14000 family of international standards. ISO, Geneva, (ISBN 978-92-67-10500-0)
ISO (2012a) Who develops ISO-standards. http://www.iso.org/iso/home/standards_development/who-develops-iso-standards.htm. Accessed 17 Jul 2012
ISO (2012b) Standards development. http://www.iso.org/iso/home/standards_development.htm. Accessed 17 Jul 2012
ISO 14025 (2006) Environmental labels and declarations, Type III—environmental declarations—principles and procedures. Geneva
ISO 14040 (1997) Environmental management—life cycle assessment—principles and framework. Geneva
ISO 14040 (2006) Environmental management—life cycle assessment—principles and framework. Geneva
ISO 14041 (1998) Environmental management—life cycle assessment—goal and scope definition and inventory analysis. Geneva
ISO 14042 (2000) Environmental management—life cycle assessment—life cycle impact assessment. Geneva
ISO 14043 (2000) Environmental management—life cycle assessment—life cycle interpretation. Geneva
ISO 14044 (2006) Environmental management—life cycle assessment—requirements and guidelines. Geneva
ISO 14045 (2012) Environmental management—eco-efficiency assessment of product systems—principles, requirements and guidelines. Geneva
ISO 14046.CD.1 (2012) Water footprint—requirements and guidelines. Geneva
ISO/TR 14047 (2012) Environmental management—life cycle assessment—illustrative examples on how to apply ISO 14044 to impact assessment situations. Geneva
ISO/TR 14049 (2012) Environmental management—life cycle assessment—illustrative examples on how to apply ISO 14044 to goal and scope definition and inventory analysis. Geneva
ISO/TS 14048 (2002) Environmental management—life cycle assessment—data documentation format. Geneva
ISO/TS 14067 (2013) Carbon footprint of products—requirements and guidelines for quantification and communication. Geneva
Marsmann M (1997) ISO 14040—the first project. Int J Life Cycle Assess 2(3):122–123
Marsmann M (2000) The ISO 14040 family. Int J Life Cycle Assess 5(6):317–318
Marsmann M, Klüppel H-J, Saur K (1997) Foreword—development of life cycle thinking. Special Issue: current LCA-ISO activities. Int J Life Cycle Ass 2(1):2–4
PAS 2050 (2011) Specification for the assessment of the life cycle greenhouse gas emissions of goods and services. BSI, London
SETAC—Society of Environmental Toxicology and Chemistry (1993) Guidelines for life-cycle assessment: a 'Code of Practice'. From the SETAC Workshop held at Sesimbra, Portugal, 31 March–3 April1993, 1st edn. Brussels and Pensacola, August 1993

Chapter 4
The UNEP/SETAC Life Cycle Initiative

Guido Sonnemann and Sonia Valdivia

Abstract The activities of the UNEP/SETAC Life Cycle Initiative have been crucial for the dissemination of LCA worldwide and the creation of a global life cycle community, since 2002, after the ISO 14040 series had been established. The Life Cycle Initiative not only contributed to capability development and the set up of national and regional life cycle networks in different parts of the world but also to enhancing and building global consensus on life cycle methodologies in areas such as life cycle inventory (LCI), life cycle impact assessment (LCIA), social life cycle assessment as well as carbon and water footprinting. Moreover, the Life Cycle Initiative has successfully promoted the way leading companies are doing life cycle management (LCM) and the integration of the three dimensions of sustainable development in life cycle sustainability assessment (LCSA).

The chapter opens with an introduction on how the UNEP/SETAC Life Cycle Initiative came to life and developed over the years. Then the special relationship of the Life Cycle Initiative to The International Journal of Life Cycle Assessment is highlighted. This section is followed by a description of the main contributions of the Life Cycle Initiative to the international community. In the subsequent section the key messages based on the work conducted during the last 10 years are described. The chapter finishes with some thoughts on the future of life cycle thinking and an introduction to the Phase 3 of the Life Cycle Initiative.

The chapter includes the list of key achievements of the Life Cycle Initiative's Phase 1 and 2 activities that are the creation of a global life cycle community, the LCI Registry, the LCIA Midpoint-damage Framework, the USEtox model, the LCM guide and capability maturity framework, the social LCA guidelines, the LCSA framework and the global guidance principles for LCA databases.

G. Sonnemann (✉)
Project Management Office of the UNEP/SETAC Life Cycle Initiative, Université Bordeaux 1, ISM—CyVi, 351 Cours de la Libération,
33405, Talence Cedex, France
e-mail: guido.sonnemann@gmail.com

S. Valdivia
Secretariat of the UNEP/SETAC Life Cycle Initiative, United Nations Environment Programme, DTIE—SCP, 15 rue Milan,
75441, Paris Cedex 09, France

W. Klöpffer (ed.), *Background and Future Prospects in Life Cycle Assessment,*
LCA Compendium – The Complete World of Life Cycle Assessment,
DOI 10.1007/978-94-017-8697-3_4, © Springer Science+Business Media Dordrecht 2014

Keywords Global capability development · LCA databases · LCI registry · LCIA midpoint-damage framework · Life cycle assessment (LCA) · Life cycle impact assessment (LCIA) methodologies · Life cycle inventory analysis (LCI) · Life cycle management (LCM) · Life cycle sustainability assessment (LCSA) · Life cycle thinking · Product sustainability information · Society of Environmental Toxicology and Chemistry (SETAC) · Sustainable consumption and production · United Nations Environment Programme (UNEP) · USEtox model

1 Introduction

In 2002 the United Nations Environment Programme (UNEP), the Society of Environmental Toxicology and Chemistry (SETAC) and partners from governments, academia, civil society, business and industry joined forces to promote life cycle approaches worldwide as a way to increase resource-efficiency and to accelerate a transition towards more sustainable consumption and production patterns. Sustainable development objectives and a company's bottom line come together in the important topic of assessing and managing the life cycle of processes, materials, products and services. After the publication of the ISO 14040 standard dealing with LCA (ISO 14040 1997), UNEP and SETAC, aware of the need for dissemination and implementation, jointly began to engage more partners to work on the articulation of science-based existing efforts around life cycle thinking and established the UNEP/SETAC Life Cycle Initiative (Life Cycle Initiative).

We would like to acknowledge here the crucial role of Helias Udo de Haes, founder and former scientific director of CML, Leiden University, The Netherlands. As chairman of the LCA Steering Committee of SETAC, he took the initiative for the establishment of the UNEP/SETAC Life Cycle Initiative, of which he has been the first and only director until 2006. Olivier Joillet, who at that time was at EPFL in Switzerland, assisted him (Udo de Haes et al. 2002).

UNEP's former Assistant Executive Director Jacqueline Aloisi de Larderel welcomed their efforts and asked them to also engage experts from Asia, North America and Latin America. In this way, Atsushi Inaba from Japan, Jim Fava from SETAC North America and Ana Quiros from Costa Rica were invited to join the organizing committee. Jacqueline Aloisi de Larderel facilitated, jointly with the former UNEP staff Bas de Leeuw and Anne Solgaard, the launch of the initiative in Prague in 2002 and the arrival of Guido Sonnemann to serve the Secretariat of the Initiative. The latter was further strengthened by the recruitment of Sonia Valdivia in 2005.

The life cycle partnership for a more sustainable world between UNEP, SETAC and public/private sector partners has the overall objective of promoting, assisting and supporting the application of life cycle thinking and life cycle approaches, including life cycle management, life cycle assessment, carbon footprinting and water footprinting, by governments as well as companies and their suppliers, customers and other value-chain partners worldwide. The final purpose is furthering sustainable innovation and global use of more sustainable products.

The Life Cycle Initiative is a response to the call from governments for a life cycle economy in the Malmö Ministerial Declaration (2000). It contributes to the 10-Year Framework of Programmes (10YFP) on Sustainable Consumption and Production (SCP), which is a process setup to promote sustainable consumption and production patterns. The 10YFP was adopted during the so-called Rio+20 World Summit on Sustainable Development in June 2012. The Initiative facilitates the exchange of knowledge of currently over 2000 experts worldwide and four regional networks from different continents.

The Life Cycle Initiative's activities to date have been carried out in two phases, in which around 200 members of the global life cycle community have been actively involved.

The first phase (2002–2007) focused on establishing the Life Cycle Initiative as a global focal point of life cycle-related knowledge and activities and on building an expert community of practitioners. Activities to move the Life Cycle agenda forward concentrated on three important fields of work:

1. Life Cycle Management (LCM),
2. Life Cycle Inventory (LCI), and
3. Life Cycle Impact Assessment (LCIA) as well as the crosscutting area of social impacts along the life cycle.

The Life Cycle Management field was added to the LCA areas of LCI and LCIA after the successful first LCM conference and related workshop organized by Allan Astrup Jensen in Copenhagen in 2001. It was considered important by UNEP to focus not only on assessment but also on the use of the life cycle approach and related knowledge in business practice (Sonnemann et al. 2001).

At the end of the first phase a process was started to help the creation of regional and national life cycle networks, in particular in developing countries, to support capability development. In particular due to the important personal engagement of Greg Norris, Harvard School of Public Health, it was possible to get life cycle networks in Africa and Latin America off the ground (Sonnemann 2004b).

Phase 2 activities (2007–2012) saw the Life Cycle Initiative evolve to be more participative with regard to stakeholders, encouraging more involvement from key actors at the global level in order to achieve common understanding and agreement on tools and strategies being developed. The main outcomes of Phase 2 were accomplished through close collaboration with crucial stakeholders in the field.

In both phases, the Life Cycle Initiative was able to provide support in the application of sustainability-driven life cycle approaches based on lessons learned from leading organizations by its capacity of engaging with world class experts and practitioners working in product policy, management and development.

The International Life Cycle Panel (ILCP) became the International Life Cycle Initiative Board (ILCB) in Phase 2 and oversaw the activities of the Life Cycle Initiative in all these years. The name change from ILCP to ICLB was done to emphasise on its actual role as governing body of the Initiative, bringing together all key partners convened by UNEP and SETAC, and to differentiate it from the UNEP International Resource Panel (IRP) launched in 2007. The ILCB plays a key role in the decision making process and provides strategic direction to the overall work of the Life Cycle Initiative. The Secretariat is hosted at UNEP and helps in implementing the tasks as recommended by the ILCB. The director and the team of programme managers in Phase 1 and the Coordination Committee headed by Jim Fava and consisting of work area chairs in Phase 2 supported the Secretariat.

In 2011 Guido Sonnemann was asked by the ILCB to lead, jointly with Bruce Vigon from SETAC, the strategy development for the next phase of the Life Cycle Initiative that was launched at the global level in Yokohama in November 2012.

In **Phase 3**, a Project Management Group co-chaired by Jim Fava and Guido Sonnemann assists the Secretariat. The vision, mission and activities foreseen in this new phase of the Life Cycle Initiative are explained in Sect. 5.

Before addressing the future of life cycle thinking in Sect. 5 as well as conclusions and perspectives in Sect. 6, the special relationship between the UNEP/SETAC Life Cycle Initiative and The International Journal of Life Cycle Assessment will be described in Sect. 2. Main contributions of the Life Cycle Initiative to the international community in Phases 1 and 2 from 2002 to 2012 will be defined in Sect. 3, and key messages based on the work conducted during the last 10 years will be highlighted in Sect. 4.

2 The UNEP/SETAC Life Cycle Initiative and The International Journal of Life Cycle Assessment

The International Journal of Life Cycle Assessment is the first journal devoted entirely to Life Cycle Assessment. It is a forum for

- scientists developing Life Cycle Assessment and Life Cycle Management,
- LCA and LCM practitioners, consultants and managers concerned about the environmental aspects of products,
- governmental environmental agencies responsible for product quality,
- scientific and industrial societies involved in LCA development, and
- environmental institutions and bodies.

That means that the target audience is similar to the one of the Life Cycle Initiative.

Due to the complementarity of the journal and the Initiative, the board of the UNEP/SETAC Life Cycle Initiative decided in 2003 to establish an official collaboration with The International Journal of Life Cycle Assessment, which became the Associated Journal of the UNEP/SETAC Life Cycle Initiative. The co-chairs of the ILCP, Jacqueline Aloisi de Larderel and Jim Fava, appreciated the efforts undertaken by the journal to globalize the use of LCA by being at that time also the official organ of the LCA Society of Japan, the Indian Society for LCA, the Korean Society for LCA, and the Australian LCA Society (Aloisi de Larderel and Fava 2003).

As part of the collaboration, the journal agreed to regularly inform about recent developments and activities of the Life Cycle Initiative and to provide active members of the Initiative from developing countries the journal for a reduced fee.

Already before this agreement, the journal reported with a special issue about the launch of the UNEP/SETAC Life Cycle Initiative. The launch took place on 28 April 2002 during UNEP's 7th High-level Seminar on Cleaner Production, and in presence of the former SETAC President Lorraine Maltby and UNEP's Executive Director Klaus Töpfer. The latter prepared an editorial for the journal and thanked its editor-in-chief, Walter Klöpffer, not only for his valuable work in promoting Life Cycle Assessment and Life Cycle Management on an international level but also for his support of the Life Cycle Initiative by this special issue (Töpfer 2002).

Since 2003 The International Journal of Life Cycle Assessment has been reporting on the Initiative's activities in the so-called Corner of the UNEP/SETAC Life Cycle Initiative in a continuous way: for example in 2005 about progresses in Life Cycle Impact Assessment within the UNEP/SETAC Life Cycle Initiative (Jolliet et al. 2005), in 2007 about the first Phase 2 activities of the Initiative (Sonnemann and Valdivia (2007) and in 2011 about the process on global guidance for LCA databases (Sonnemann et al. 2011).

Furthermore, it has published relevant deliverables such as the LCIA Midpoint-damage Framework of the UNEP/SETAC Life Cycle Initiative in 2004 (Jolliet et al. 2004), the activity of Task Force 1 on global life cycle inventory data resource (Curran 2006) and a special issue on USEtox in 2011 (Hauschild et al. 2011). The secretariat has been negotiating with the journal an open access to a number of those deliverables.

The journal has also been helpful in announcing conferences such as CILCA (International Conference on Life Cycle Assessment) in Costa Rica in 2005 (Sonnemann et al. 2005)[1] and the recent Indian life cycle assessment and management conference in 2012 (Datta et al. 2012) as well as in reporting on events such as in the form of key observations arising from papers on sustainable production, use and recycling of natural resources from the symposium in Portland in 2006 (Fava et al. 2006).

We expect this fruitful cooperation for enhancing the state of worldwide LCA development to continue in the future. As a first step updates on recent developments in Life Cycle Impact Assessment and the finalization and current dissemination activities of the publication on global guidance principles on LCA databases are foreseen. Moreover, special issues on Life Cycle Sustainability Assessment and global land use impacts on biodiversity and ecosystem services in LCA are under preparation.

[1] The conference series provides an international forum to share experiences on Life Cycle Thinking and related tools. CILCA is a bi-annual event which is held in different countries of Latin America convoking experts and interested audiences from across the globe. The first CILCA was held in 2005 in San José Costa Rica, and from there has followed a successful journey which included locations spread all along the region: Sao Paulo (Brazil) in 2007, Pucón (Chile) in 2009, and Coatzacoalcos (Mexico) in 2011.

3 Main Contributions from 2002 to 2012 of the Life Cycle Initiative to the International Community and Best Examples Worldwide

In this section we will highlight the main contributions from 2002 through 2012 of the Life Cycle Initiative to the international community. This includes relevant deliverables as best examples worldwide. Those deliverables that have been published in official UNEP and SETAC documents, have got an ISBN number and are parts of scientific journals, in particular The International Journal of Life Cycle Assessment, are referenced throughout the text. Other reports, training material and tools mentioned are available at the website of the Life Cycle Initiative at http://www.lifecycleinitiative.org.

3.1 Phase 1—Creating a Global Community

In Phase 1 from 2002 to 2007, programmes to move the life cycle agenda forward concentrated on three important fields of work:

1. Life cycle management,
2. Life cycle inventory, and
3. Life cycle impact assessment including the crosscutting area of social impacts along the life cycle.

3.1.1 The Life Cycle Management Programme

The Life Cycle Management programme was oriented to the application of life cycle approaches aiming to bring LCA and life cycle thinking into the **practice of business** and into policy decision-making. The specific aims were to:

- create awareness and improve skills of decision makers by establishing forums for best practice and carrying out training programmes all over the world;
- document experiences on practical applications of Life Cycle Thinking and to highlight enablers and barriers for development and implementation of a LCM approach;
- develop a LCM framework for different tools and concepts, including e.g. Integrated Product Policy or Extended Product Responsibility;
- take into account economic, social and ethical aspects, and occupational health and safety, risk management, community outreach and other related programs;
- identify needs and availability of training modules and dissemination;
- make recommendations on how to approach the needs of developed, emerging and developing economies as well as SMEs.

Deliverables from this programme that partially continued beyond 2007 include:

- LCM definition study;
- LCM background document;
- Life Cycle Management—A business guide to sustainability (UNEP/SETAC 2007);
- Communication of life cycle information in the building and energy sectors (UNEP/SETAC 2008);
- LCM training kit material in English, French, Spanish and Portuguese.

The following Task Forces (TFs) had been established under the Life Cycle Management (LCM) Programme to help achieving this deliverables:

- LCM Handbook (LCM TF 1)
  - In absence of an agreed upon definition for LCM, the handbook was aimed to introduce the LCM framework, discuss drivers and needs, describe the major underlying approaches and provides selected illustrative examples and successes while introducing and using LCM in practice. As part of the deliverables of this Task Force, a Training Kit on LCM for Trainers and Delegates had been developed.
- Life cycle based product development (LCM TF 2)
  - The integration of environmental considerations along the products life cycle in the product development process, rather than as an isolated function, was subject of this task force. Results of its discussions served the development of the LCM Handbook by the LCM TF2.
- Communication of life cycle information (LCM TF 3)
  - The task force had the aim to position the existing tools for the communication of life cycle information and identify the best options to initiate changes of consumption and production patterns. In particular the task force examined the mutual reinforcement amongst the tools and within the larger LCM framework, specifically within management systems.
- Management and Stakeholder engagement along the life cycle (LCM TF4)
  - Management along the life cycle is to approach and apply life cycle thinking from the management system point of view by using the ISO14001 and 14004 but also other standards such as ISO14031 on indicators together with GRI indicators (for environmental reporting!). Management along the life cycle can be seen as check lists of potential environmental and other sustainability aspects to be included in a life cycle oriented management system (sometimes referred as Product-Oriented Environmental Management System).

3.1.2 The Life Cycle Inventory Programme

The Life Cycle Inventory programme refers to the second phase of LCA and aimed at increasing the access to and quality of LCI databases. The specific aims of the programme were to:

- Identify user needs for data, and needed/desired data characteristics;
- Identify user needs for further LCI methodological guidance and consistency;
- Increase the capacity for making, and judging/validating, LCIs globally;
- Provide users of LCI data the broadest possible view of available LCI data options and the consequences of data selection for results quality and validity;
- Develop and put into place mechanisms or processes that stimulate continuous improvement in data availability, quality, and transparency.

Deliverables stemming from the Life Cycle Inventory programme include:

- LCI definition study,
- Report on Activity of Task Force 1: Data Registry—Global Life Cycle Inventory Data Resources (Curran 2006),
- LCI Database Registry,
- LCI Format Converter,
- Report for Task Force 3: Inventory methods in LCA—towards improved methodological consistency (Lundie et al. 2007),
- Initiation of national and regional life cycle networks (Sonnemann 2004b).

The following Task Forces were established under the Life Cycle Inventory (LCI) Programme:

- LCI Database Registry (LCI TF 1) and LCI Database Characteristics and Quality (LCI TF 2)
  - Task Forces were responsible for developing the UNEP/SETAC Database Registry: a comprehensive, web-based listing of available LCI databases for the world LCA community.
  - The LCI TF 1–2 pursued consistency on four core characteristics of databases:

  Data quality,
  Documentation format,
  Data exchange format,
  Nomenclature.

- LCI Methodological Consistency (LCI TF 3)
  - The aim of this task force was to initiate and stimulate processes, studies, and forums that facilitate voluntary and practice-oriented movement towards transparency, ultimately contributing to improved consistency and commonality of LCI methodological practice.
- LCI Databases and Capacity Building (LCI TF 5)
  - Task Force 5 was a practical one involving many regions (teams from novice and experts) and aiming at

  Practical involvement,
  Job training (training the trainers),
  Capacity building,
  Operational 'field tests' for TF 2 and 3 and I/0 or hybrid LCI,
  Assisting in finding funding.

The first three Task Forces were orientated towards experience sharing and guidance, while the last Task Force was training-orientated supporting the capacity building efforts of the Life Cycle Initiative.

3.1.3 The Life Cycle Impact Assessment Programme

The Life Cycle Impact Assessment programme refers to the third phase of LCA and dealt with the evaluation of environmental impacts, (e.g. climate change and toxicity) of products and services over their whole life cycle. The aim of the LCIA programme was to increase the quality and global reach of the life cycle indicators by promoting the exchange of views among experts. Its specific aims were to:

- Identify user needs for Life Cycle Impact Assessment;
- Provide a clear picture of the impact categories, including different impacts than the one typically applied in "OECD country lcas", like e.g. Erosion or biodiversity;
- Provide guidelines for the starting points, the decision-making framework and guidelines for the identification of recommended practice
- Identify case studies, and industrial partners, to test and improve the method feasibility;
- Identify the links with the LCI and LCM programmes, including the relation of LCIA to indicators, which also include the economical and social dimensions of sustainability.

Key deliverables prepared in this programme are the following:

- LCIA definition study;
- UNEP publication: Evaluation of Environmental Impacts in Life Cycle Assessment (UNEP 2003);
- A paper on the LCIA Midpoint-damage Framework of the UNEP/SETAC Life Cycle Initiative (Jolliet et al. 2004);
- A paper on the progresses in Life Cycle Impact Assessment made within the UNEP/SETAC Life Cycle Initiative (Jolliet et al. 2005);
- SETAC publication: Life-cycle assessment of metals—issues and research directions (SETAC 2005);
- A scientific paper on the key elements in a framework for land use impact assessment within LCA (Mila i Canals et al. 2007);
- Declaration of Apeldoorn on Life Cycle Assessment of Non-Ferrous Metals and related scientific articles such as on the Clearwater consensus for the estimation of metal hazard in fresh water (Diamond et al. 2010);
- USEtox model as an environmental model for characterization of human and ecotoxic impacts in LCIA and for comparative assessment and ranking of chemicals according to their inherent hazard characteristics (Rosenbaum et al. 2008);

- A report on guidance on how to move from current practice to recommended practice in Life Cycle Impact Assessment, in particular for transboundary impacts.

The Life Cycle Impact Assessment programme was established with the following four task forces:

- LCIA information system (LCIA TF 1)
  - Towards the enhancement of the availability of sound LCIA data and methods, this Task Force aimed to develop an LCIA information system and to finalize and extend the general framework.
- Natural resources and land use (LCIA TF 2)
  - This task force aimed at establishing recommended practice and guidance for natural resources and land use categories, i.e.: water resources, minerals resources, energy carriers, soil resources and erosion, land use, salinisation and desiccation and biotic resources. It addressed both midpoint categories and their relation to damage categories such as the biotic and abiotic natural environment.
- Toxicity impacts (LCIA TF 3)
  - Identification and quantification of impacts on human health and on ecosystems linked to the use and emissions of toxic substances were of central importance to the development of sustainable technology. On the one hand, the UNEP/SETAC Life Cycle Initiative made use of significant recent progress in LCIA of toxics. On the other hand, several crucial shortages of present methodologies were addressed to enable a proper interpretation of LCI results.
- Transboundary impacts (LCIA TF 4)
  - This task force aimed at establishing recommended practice and guidance for use in transboundary categories, i.e.: climate change, ozone depletion, aquatic and terrestrial eutrophication and acidification, photooxidant formation and respiratory inorganics.

3.1.4 Crosscutting Activities

The aim of the crosscutting activities was to address questions in relation to life cycle approaches that were identified as relevant in the overall user needs assessment, but that have not been further developed as part of the Definition Studies since they included topics that concern more than one programme.

Main outcomes of the crosscutting activities were the following deliverables:

- Report 'Life Cycle Approaches—The road from analysis to practice',
- Brochure 'Why Take a Life Cycle Approach?' (UNEP/SETAC 2004, translated into French, Spanish, Chinese and Japanese),
- Feasibility Study on the integration of social aspects into LCA,

- Life Cycle Management navigator for SMEs (prepared jointly with the LCM programme).

The life cycle based topics embraced under the heading of crosscutting activities were (Sonnemann 2004a):

- Simple life cycle based tools,
- Integrated resource and waste management,
- Integration of social aspects into LCA,
- Function-Based Approach.

While TFs 1, 2, 3 were experience-sharing- and guidance-orientated, the focus of TF 4 was practice and training-orientated and closely linked to the task force on LCI Databases and Capacity Building. The work on integrated resource and waste management was taken over by the International expert group on Life Cycle assessment for integrated waste management (Coleman et al. 2003) and the International Resource Panel (IRP).

An overview of Phase 1 structure is given in Fig. 4.1.

3.2 Phase 2—Becoming a Stakeholder

3.2.1 Overall Structure

In 2006, the strategy for a new phase was developed to give a new focus and ensure that on-going activities were finalized as far as possible in a given period. Key activities like the promotion of Life Cycle Management, the development of an LCI registry and the USEtox model continued in Phase 2. The achievements with regard to the national and regional networks and the establishment of a truly global life cycle community were used to foster capability development on life cycle approaches worldwide among other activities through the launch of an LCA award for developing country projects and the support of the organization of conferences in emerging economies such as Brazil, China, India, Mexico and South Africa.

The identified objectives for Phase 2 were met through projects in five Work Area Interest Groups (WAIG), as indicated in Fig. 4.2:

A. Life Cycle Approaches for Methodologies and Data (including data, methods, case studies, etc.);
B. Life Cycle Approaches for Resources and Impacts (including natural resources, chemicals, water, energy, etc.);
C. Life Cycle Approaches for Consumption Clusters (structured in housing, mobility, food and consumer products);
D. Life Cycle Approaches for Capability Development (including institutional empowerment, training, curricular development, etc.;
E. Life Cycle Management in Businesses and Industries.

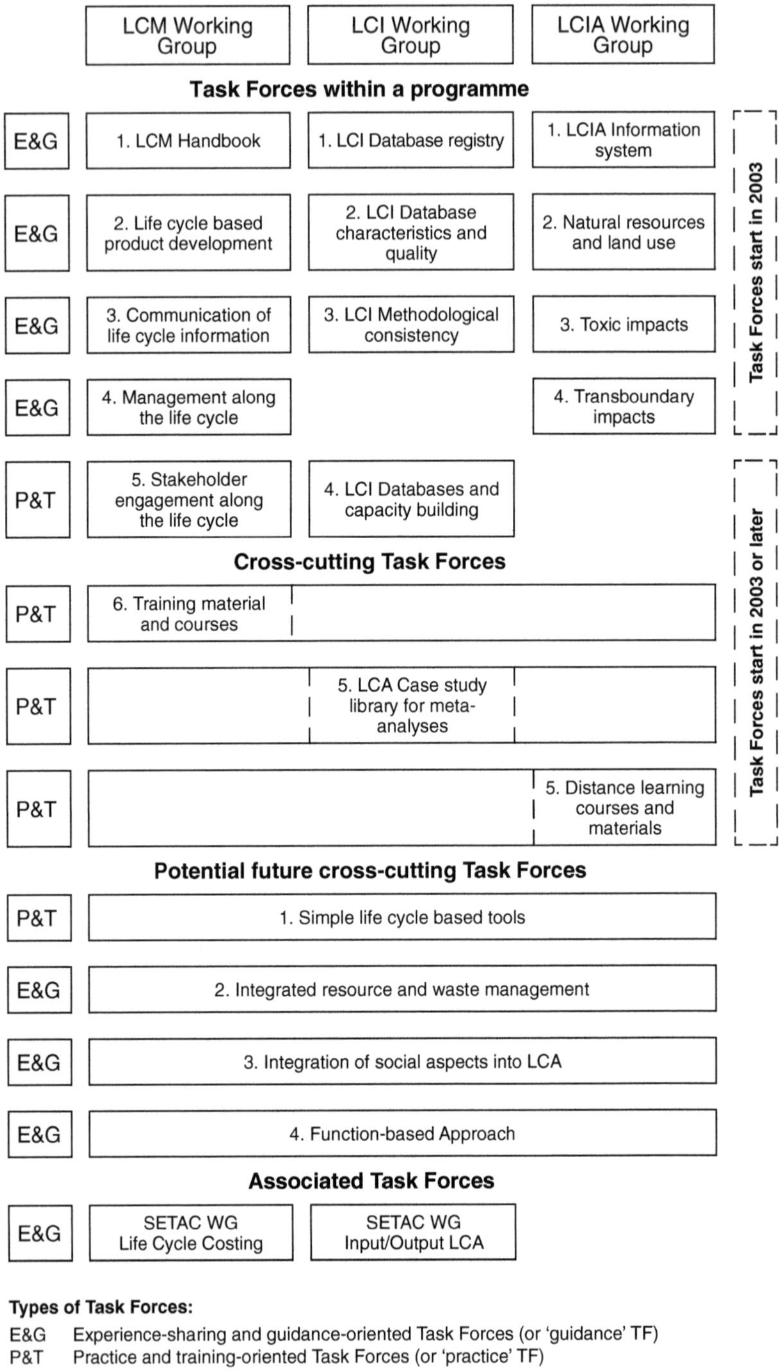

Fig. 4.1 Overview of working groups and task forces in phase 1. (Sonnemann 2003)

Fig. 4.2 Relationship among the five work area interest groups chosen for the practical application approach in phase 2. (UNEP/SETAC 2012b)

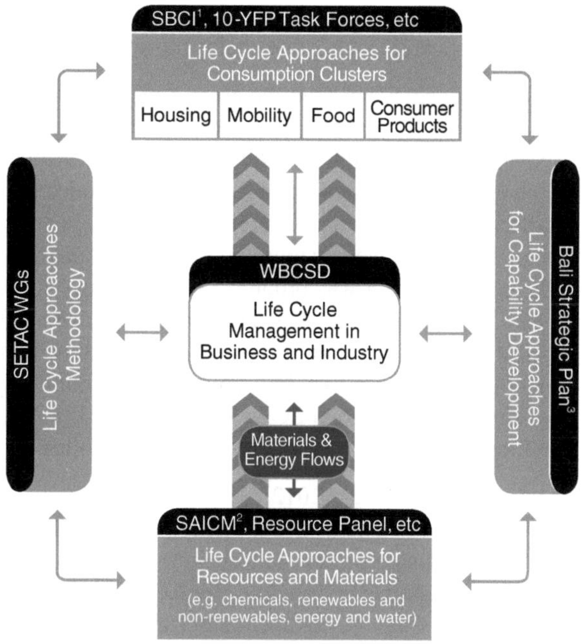

Figure 4.2 indicates the relationship among the WAIGs. In this way it was expected that the impact of the Life Cycle Initiative goes, beyond the work on methodologies and capacity building, to practical applications that make a difference in the real world and thus contribute more effectively to the on-going international efforts to change unsustainable patterns of consumption and production. The expectations for each of the objectives were met by having the Secretariat conducting the work and the experts carrying out projects in the five Work Area Interest Groups.

3.2.2 Deliverables

Relevant UNEP publications as products of the Phase 2 activities are:

- Greening the Economy through Life Cycle Thinking—10 Years of the UNEP/SETAC Life Cycle Initiative (UNEP/SETAC 2012a),
- Global Guidance Principles for Life Cycle Assessment Databases—A Basis for Greener Processes and Products (UNEP/SETAC 2011a),
- Towards a Life Cycle Sustainability Assessment—Making informed choices on products (UNEP/SETAC 2011b),
- Guidelines for Social Life Cycle Assessment of Products (UNEP/SETAC 2009a, translated into French and Dutch),

- Life Cycle Management—How business uses it to decrease footprint, create opportunities and make value chains more sustainable (UNEP/SETAC 2009b).

A particular achievement in the area of LCM is the Life Cycle Management capability maturity framework. The latter shifts the focus from driving performance on prescriptive sustainability metrics to building the capacity of organizations in a supply chain to identify and manage social and environmental issues in a manner that is tailored to their business strategy. It helps the supplier to identify where and how to start and continue their journey towards sustainability (Swarr 2011).

A related product was produced by a SETAC working group: the Environmental Life Cycle Costing: A Code of Practice (SETAC 2011). Thanks to this SETAC publication it was possible to present a common framework for Life Cycle Sustainability Assessment (LCSA), covering environmental LCA (E-LCA), Life Cycle Costing (LCC) and social LCA (S-LCA) (UNEP/SETAC 2011b).

Moreover, training kits and courses on the following topics were developed:

- Water Footprinting—2012,
- Life Cycle Management Capability Maturity Model: helping SMEs apply
  - LCA in business decision-making—2012,
  - Global Guidance Principles for LCA Databases—2012,
- Social Life Cycle Assessment and Life Cycle Sustainability Assessment—2011,
- LCA (Life Cycle Assessment) Training Kit Material—2008.

Finally, project groups have organised an important number of workshops and published relevant supporting documentation and scientific papers in international journals:

- WULCA, the project group on the Assessment of Use and Depletion of Water Resources within LCA, has provided relevant input through the SETAC liaison role, to the ISO standardisation process on water footprintng (ISO 2012) and has published articles such as a framework for assessing off-stream freshwater use in LCA. (Bayart et al. 2010);
- The project group on Carbon Footprinting has managed to provide technical input to the WBCSD/WRI Greenhouse Gas Protocol WBCSD/WRI (2011) and the ISO standardisation process on carbon footprinting (ISO 2013), ensuring that both standards are based on ISO 14040 and close to each other (Finkbeiner 2009);
- The project group on Integrating Human Indoor Air Pollutant Exposure within Life Cycle Impact Assessment has proposed a new methodological framework for a general procedure to include human-health effects from indoor exposure in LCA (Hellweg et al. 2009);
- The project group on Global Land Use Impacts on Biodiversity and Ecosystem Services in LCA is working towards widely accepted characterisation factors. (Koellner et al. 2013);
- Methodological sheets for social LCA have been developed by the project group working on social and socio-economic LCA (Benoît-Norris et al. 2011).

3.2.3 Running a Multi-Stakeholder Process: Global Guidance for LCA Databases

A particular challenge in the second phase was to run the process on 'global guidance for LCA databases' towards overall consensus since a number of stakeholders had controversial views in this field of LCA. The process was launched at the first Stakeholder Engagement Meeting, 'Towards Global Guidance for LCA Databases', in Boston in September 2009, where the high attendance confirmed the international interest in the UNEP/SETAC proposal. Many participants felt that the process was very timely and a majority of the participants agreed with the vision, which was to help provide global guidance on the establishment and maintenance of LCA databases, as an input for improved interlinkages of databases worldwide. The vision was expected to contribute to increasing the credibility of existing LCA data, to further foster the generation of more data (also for applications such as carbon and water footprint) and to enhance their overall accessibility.

The process was overseen by a Steering Committee consisting of stakeholders from governments, industry and academia/NGOs (Sonnemann et al. 2011). In seven stakeholder meetings following the launch the interested audience was informed about the plan for the development of a global guidance. The central activity was a 5-day Pellston-type Workshop in Shonan Village in January/February 2011 organized by the Secretariat of the Life Cycle Initiative on behalf of UNEP and SETAC, in close co-operation with the Japanese hosts.

The workshop participants included selected experts from on-going regional and national as well as industry database initiatives in OECD countries, emerging economies and developing countries. Moreover, a few key consultants developing databases as well as experienced SETAC and regional life cycle network experts were also attending together with UNEP staff and relevant users of LCA databases. The workshop participants were able to put together the basis for a publication on the Global Guidance Principles for LCA Databases, called Shonan Principles, which was launched in Berlin in August 2012. These principles give guidance for proper gathering and management of data, which enable better, more reliable life cycle assessment results and improve their use for decision-making. (UNEP/SETAC 2011a)

4 Key Messages Based on Work Conducted During the Last 10 Years[2]

The UNEP/SETAC Life Cycle Initiative members and its network of stakeholders and professionals in the field believe that the transition to a green economy can only be successfully accomplished if the decisions made toward this goal are based

[2] This section is reproduced from the publication of UNEP/SETAC (2012) Greening the Economy through Life Cycle Thinking—10 Years of the UNEP/SETAC Life Cycle Initiative, 60 pp, Paris, France, with the authorization of UNEP as copyright holder.

upon solid, science-backed information. Life cycle thinking, through its many approaches and tools, helps to identify both the negative and positive consequences of decision-making to the sustainability triple-bottom line, thus enabling an appropriate weighing of options.

In support of this position, the Life Cycle Initiative has developed the following seven key messages. The messages describe the current state of use of life cycle approaches and tools, their beneficial qualities that support the transition to a green economy, and a glimpse into the future of life cycle approaches and tools.

4.1 Life Cycle Thinking in the Private Sector—Ahead of the Curve

Many companies, both large and small have realized that introducing sustainability into operations management makes business sense. Management trends today are now moving towards a true triple-bottom line approach, supported by a mature, broad spectrum of life cycle approaches and tools that can be selected and tailored to examine specific issues or impact areas, and are supported by knowledgeable networks of stakeholders and professionals, reliable data and standardized methodologies. Using this approach, the ultimate goal of addressing the environmental impact of a product over its life cycle has changed. Before, the goal was to make it less damaging, whereas now, a potential goal is that it leads to an improvement of the environment. In addition, the application of life cycle tools has been extended to simultaneously consider social and economic aspects, thus providing an approach to measure changes to societal well-being and wealth.

Life cycle thinking that influences product design, strategic planning, procurement, and sales helps businesses:

- Enhance their image and the value of their brands—businesses can avoid criticism since responsibilities are better defined along the supply chain. Suppliers are made responsible for their own share of impacts as well as for corrective and preventive actions;
- Find new ways for marketing and sales departments to communicate and interact with customers—a company can promote its products and services supported by positive social and environmental claims derived from an impartial and science-based approach;
- Share life cycle information with suppliers, customers, and waste handlers to identify risks and opportunities for improvement—the risks might relate to the environment, human health, safety, or finance. Opportunities here include increasing market share, improved brand image, more effective use of materials, and innovation, amongst others.

The private sector is incorporating life cycle thinking on many fronts, including:

- Product development (via design for environment, design for disassembly…);

- Production (via environmental and social life cycle assessment, carbon footprint, water footprint, material flow accounting, supplier codes of conduct, supplier audits…);
- Marketing (via use of eco-labels, social and environmental certifications and labels, environmental and/or social product declarations…);
- Use (via demand-side management, integration of monitoring and communication technologies into products to inform users on energy and water use, design for easy maintenance, product service system development, user training on environmentally friendly product use…);
- Disposal (via design for disassembly or compostability, participation in product recycling systems or product take-back programmes…);
- Management (via operational transparency through reporting on environmental, social, and sustainability performance; corporate social and environmental responsibility; commitment to continuous improvement…).

For the most part, companies start with the use of some life cycle approaches and tools on a product-oriented or project-organized basis. For example, in Asia and Latin America the need and use of water and carbon footprints is increasing exponentially.

The next step is to broaden the integration of life cycle thinking on a 'top to bottom' basis, including internal policies, management systems, accountabilities, and incentives—and at the same time, applying these elements wherever possible to yield improvements across the value chain.

There are some companies that are forging ahead by working with suppliers and supply chain issues towards continuous improvement as an important strategic consideration. Realizing that their future relies on "sustained profits", these companies are taking bold steps forward to fully address the triple bottom line of sustainability.

Over the past decade of activities, the Life Cycle Initiative has helped to improve the understanding in the private sector of the benefits that can be derived from implementing life cycle methodologies and using the related tools within an environmental management framework. Workshops, study trips, and seminars brought together experts and stakeholders from many industry sectors (e.g. automotive, forestry, building and construction, electronics…) to share experience, and move the life cycle agenda forward in those sectors.

The Life Cycle Initiative produced several reports, guidance document and training materials and courses touching on most of the elements relevant to Life Cycle Management to support the uptake of life cycle approaches and the use of life cycle tools in the private sector.

4.2 Life Cycle Thinking in the Public Sector—Potential for Improvement

Life cycle approaches are not new to the public sector, particularly in industrialised countries. Life cycle costing was used in the 1960's by the US army to assess the full life cycle costs of investments in tanks and tractors. Since then, life cycle approaches and tools have been making inroads to address the environmental, and even the social side of government operations.

Public spending normally represents 8–30 % of national Gross Domestic Product and every purchase is an opportunity to drive markets towards innovation and sustainability. Purchasing products and services that are "environmentally preferable" reduces the impact government operations have on the environment and supports regional and global markets for environmentally "preferable" products and services. This approach has matured since the 1990's to the extent that green and sustainable public procurement strategies at the national, provincial/state and local level have been implemented in both industrialized and developing countries alike.

Good examples of policies based on life cycle approaches are already in place. On the production side, pollution limits and cleaner production are typically supported by regulations and often by economic instruments, such as green taxes or emissions trading systems operating on a regional or international basis. On the product side, policies (in addition to sustainable public procurement noted above) have been put into place that encourage the development of green products, including ecodesign directives, or material recovery programmes. Extended producer responsibility regulations make producers responsible for their products from production through final disposal, and therefore, provide an incentive to develop products with improved environmental performance in all stages of the product life cycle. Regarding policy measures to support a shift to renewable energy, some countries encourage the installation of renewable energy capacity by offering premium feed-in tariffs for solar generated electricity.

However, these examples could be described as a dartboard approach, addressing individual issues with specific policies and tools. There are some forerunners in the public sector that have seen the benefit to be gained from going beyond the dartboard approach, toward fully integrating a life cycle perspective into all areas of government operations and policy. The bold steps taken by these public authorities will yield dividends for the sustainability of their constituencies and stakeholders, and will be the ones to watch as lessons are learned from their progress.

There are several measures that can be taken by governments at all levels to create an enabling environment for life cycle thinking and approaches to gain a foothold, and help to set the course for the transition to a green economy.

First, governments can support data gathering and information sharing on the state of the environment, ecosystems and biodiversity, as well as for social indicators. Going further, an assessment of the wide range of environmental and social impacts of upcoming policies from a life cycle perspective can help to identify priority impact areas, and provide the knowledge required to avoid decisions that may undercut environmental conservation and social well-being.

Second, when subnational or national governments design policy, negotiate voluntary agreements with industry, or decide where to invest resources, life cycle thinking can apply. Measuring potential life cycle impacts of decisions can help governments to:

- Inform government programmes and help prioritise these programmes, based on life cycle information;
- Make policies more consistent among consumers, producers, material suppliers, retailers, and waste managers and also among different policy instruments (such as harmonising regulations, voluntary agreements, taxes, and subsidies);
- Promote pricing products and services to accurately reflect the costs of environmental degradation, health problems, erosion of social welfare, and impacts at other life cycle stages. Such "price signals" can send messages to consumers and provide incentives for businesses to continuously improve the environmental and social performance of products or services, across each stage of the life cycle;
- Introduce policies that support take-back systems to establish a recycling-based economy according to the hierarchy reduces, reuse and recycle.

In support of the public sector taking on life cycle perspectives, the Life Cycle Initiative organized the Third Chinese Roundtable on Sustainable Consumption and Production with a focus on Life Cycle Assessment and Life Cycle Management in Beijing in 2009. The participants represented international organizations, Chinese national and local government agencies, Chinese companies from different industry sectors including minerals & metals, building & construction and ICT, and research institutes, from China and overseas.

Moreover, joint efforts were initiated with UNEP activities such as the Marrakech Process on Sustainable Consumption and Production, the Sustainable Building and Construction Initiative and the Strategic Approach to International Chemicals Management in order to inform decision-makers about life cycle approaches.

4.3 *Life Cycle Methodologies, Impact Assessment and Data—The Foundation for Informed Decision-Making*

Methodologies and tools are the working-level of life cycle thinking. They can be put into practice in many ways and towards many ends. For those who are new to life cycle thinking, they may be surprised to learn that many thousands of individuals use life cycle tools daily in their decision-making. Purchasing consumer goods while considering the water or energy use information provided by environmental product declarations is one way. Buying food while considering labels for certified organic produce and meat is another. Shopping for textiles and clothing while considering social and eco-labels is yet another.

Progress in making life cycle tools user-friendly with easy to interpret outputs is a result of years of data gathering and sharing, database development, refining of methodologies, and the development of appropriate means of communication. Accessibility has expanded from its debut in universities and research centres to one that is used daily by people of all ages and around the world via the internet. These on-line tools enable a personal water, energy, or ecological footprint to be derived and in so doing, allow the user to identify where the environmental (or social) impact of their consumption and lifestyle is the greatest.

The ease with which more involved assessments are undertaken has also benefitted from developments over the past decade. With the right tools, life cycle 'screens' can be completed within several hours. Quick studies can take several days. Larger assessments can take up to several months, depending on the need for new data collection. Once main operations are modelled, studies can be performed quickly.

However, with the large number of methods, labels and calculators now available, there is a risk of confusion of which is best to use for decision-making. It is therefore important that methods are harmonized to generate assessment results that are consistent, comparable, and transparent. Also, one can note certain limitations built into some tools in that the availability of locally relevant data is often limited.

The Life Cycle Initiative, with the tremendous dedication of its task force and project group members, has been able to expand the variety and at the same time increase the robustness of methodologies and tools, and consolidate international consensus around them (e.g. LCM concept, the USEtox model, social LCA guidelines, guidance principles for LCA Databases, etc.). In addition, gaps and limitations of LCA tools and techniques have been explored and addressed, also with international acceptance (e.g., life cycle impact assessment framework including climate change and ozone depletion impacts).

The deliverables have focused on life cycle management, life cycle inventory and impact assessment methodologies, databases, the impact of chemicals, the development of a social life cycle assessment framework, jointly with life cycle costing as a part of the broader life cycle sustainability assessment. All the publications, reports and scientific papers derived from these activities have been mentioned before.

4.4 *Life Cycle Sustainability Approaches—Measuring Triple Bottom Line Impacts*

In addition to tackling economic questions when developing policies and strategies, or products and services, governments and enterprises are under increasing pressure to consider impacts on the environment and society. The growing societal concern with addressing the three pillars of sustainability (i.e. environment, economic, and social) requires that appropriate tools are available to inform decision-making. Up to now, environmental LCA and life cycle costing have been applied to assess the environmental and economic aspects. The recent addition of social LCA to the life cycle toolbox puts the last piece in place for a life cycle sustainability assessment.

One key objective of the Life Cycle Initiative is to help extend LCA methodologies beyond their original scope of identifying and assessing resource consumption and environmental interventions associated with products or processes. LCA can be extended in many ways, but one major advance has been the elaboration of methods and techniques that can measure sustainability, thus allowing LCA to support decision-making toward more sustainable product and process systems. Synthesizing these methods with life cycle techniques has enabled the elaboration of a life cycle sustainability assessment (LCSA).

LCSA contributes to the discussions on sustainable development as a methodology with a great deal of potential to provide a combined sustainability indicator of a product or process by combining environmental life cycle assessment (E-LCA), social life cycle assessment (S-LCA) and life cycle costing (LCC) in a coherent. The benefits of a simultaneous assessment of the three sustainability pillars in one tool, as opposed to using three separate tools are numerous:

- Helps clarify the trade-offs between the three sustainability pillars, life cycle stages and impacts, products and possibly;
- Implies the ability to reduce environmental degradation and the use of natural resources in a cost-effective manner, while at the same time contributing to social welfare;
- Promotes awareness on triple bottom line sustainability issues in value chain actors;
- Supports enterprises and value chain actors in identifying weaknesses and enabling further improvements of a product life cycle;
- Supports decision-makers in prioritizing resources and investing them where there are more chances of positive impacts and less risk of negative ones;
- Helps decision-makers choose sustainable products and technologies; consumers will not only know which products are more cost-efficient, eco-efficient or socially responsible, but also more sustainable;
- Provides guiding principles to achieve sustainable consumption and production.

The Life Cycle Initiative led the effort to develop this framework, resulting in the publication 'Towards a Life Cycle Sustainability Assessment: Making Informed Choices on Products'. The aim of the publication is to support stakeholders looking for approaches that will provide holistic assessments of the implications of a product's life cycle for the environment and society. The publication includes eight case studies that illustrate how current and emerging life cycle assessment techniques are being implemented worldwide.

4.5 Trade-Offs and Unexpected Consequences—Avoiding the Pitfalls

It has been proven time and time again that making decisions with a limited vision of a problem can be counter-productive, and in extreme cases, even take society in the wrong direction when unexpected consequences occur.

Trade-offs will always be a part of decision-making, but when a life cycle perspective is considered, it expands the field of vision of the issue at hand. Looking up and down the value chain can help to reveal acceptable and unacceptable trade-offs, and may uncover otherwise unexpected consequences that could occur—in diverse stages of the value chain, to other sustainability pillars, to other societies, and so on. Because it is holistic, systemic and rigorous, life cycle assessment is an essential tool for generating information and broadening knowledge about potential and real impacts along a product's life cycle, and thereby increases the possibility to improve overall product sustainability.

Potential trade-offs can be characterized in many ways.

4.5.1 Trade-Offs Between Stages of the Product Value Chain

From its humble beginnings as a raw material taken from the Earth, a product and its components can travel thousands of kilometres and be handled and used by hundreds or thousands of people before it reaches its final disposal phase. Similarly, a decision to use one raw material over another can have an impact on each link of the product value chain.

For example, consider a car made with lightweight composite materials as opposed to conventional steel. While the benefits of lighter automobile weight can translate into fuel savings in the use phase, the production and disposal or recycling of composite materials need to be assessed as well and compared to conventional steel production and recycling in order to truly know which will be more beneficial to society and the environment.

4.5.2 Trade-Offs Between Environmental Impact Categories

Land, water and air are intricately involved in the human life cycle, as well as in the life cycle of products. Decisions made in the name of protecting one of these environmental 'media' can result in the detriment of another, and possibly lead to consequences for human health.

A classic example is MTBE (Methyl Tertiary Butyl Ether). MTBE is added to gasoline to increase octane levels and enhance combustion, which in turn reduces polluting emissions. MTBE in gasoline can reduce ozone precursors by 15 %, benzene emissions by 50 %, and CO emissions by 11 %. While MTBE helps mitigate air pollution, the MTBE itself may be toxic if not combusted fully. MTBE is not considered highly toxic, but it has been banned for use in many US states. Of most concern is the MTBE found in lakes, reservoirs, and groundwater for potable water supplies. Levels of MTBE in the environment are now measured when MTBE is suspected to have evaporated from gasoline or leaked from storage tanks, lines and fuelling stations.

4.5.3 Trade-Offs Between Sustainability Pillars: Environmental, Social, Economic

In a green economy based on the principle that products and services should ultimately be to the benefit of the environment and society over their life cycle, the full costs of protecting the environment and appropriate conditions and treatment of workers must be considered. In other words, a product should be produced to the detriment of neither the environment nor the people participating in the value chain.

For example, the global textile and electronics industries have come under scrutiny for producing inexpensive clothing and electronic equipment to the economic benefit of several global enterprises and their consumers, while using inappropriate labour practices that are socially detrimental to the people working in the production of these items. In a contrasting example, organic farming may not only be less damaging to the environment when compared to conventional farming methods (e.g., regarding chemical use) it can also improve farmers' working conditions and provide overall health benefits to society.

4.5.4 Trade-Offs Between Societies/Regions

In the globalized economy, product value chains are spread across countries around the globe. Decisions made to address an issue in one region can thus cause unexpected impacts in other parts of the world.

For example, with respect to electronic waste (e-waste), it could be said that "one person's waste is another person's gold" since electronics contain many valuable and recyclable materials (e.g., copper). The increasing popularity of electronic goods over the past two decades, and the rate at which new products are developed, has led to the creation of ever increasing amounts of e-waste to be recycled. However, acquiring the 'gold' from e-waste recycling has come at a high environmental and social cost in some developing countries. Directives for e-waste recovery and recycling in some industrialized countries led large amounts of e-waste to be recycled informally in developing countries under detrimental conditions for the environment and the people doing the recycling, due to releases of toxic materials in the process. New directives approved in the EU in 2012 have strengthened provisions against the exportation of e-waste.

4.5.5 Generational Trade-Offs

Sustainable development is about making decisions today that preserve the ability of future generations to meet their needs. The Native-American Ojibwe tribe recognized this, and as a principle, made their decisions considering, as much as possible, the lives of seven generations of children. In comparison, our current globalized economy, largely responsible for the state of the environment we live n today, typically considers a far shorter time span. Long- term business decisions are made for time periods of 10–20 years as a maximum, or, less than one generation.

A good example of this is the current debate over climate change and the fluctuating international commitment to reducing greenhouse gas emissions. Decisions made now are going to have an enormous impact, for better or for worse, on future generations and the stability of the climate in the future. In a contrasting example, we can see the positive result of the consideration of future generations in decisions made to address the holes forming in the Earth's ozone layer. The result of the Montreal Protocol's phase out of ozone depleting substances from consumer and industrial products has largely been seen as a success in securing the future of the ozone layer, and protecting human health and well-being for generations to come.

4.5.6 Relevant Activities in Last 10 Years

The Life Cycle Initiative engaged in developing life cycle practitioner capabilities in non-OECD countries while building and supporting global and national networks of practitioners and stakeholders. Greater national capability and knowledge is the foundation for ensuring that diverse perspectives and local information/data are considered in life cycle assessments and should allow trade-offs to be more readily identified, and unexpected consequences to be avoided.

Work with partners, such as the International Resource Panel has already resulted in work on identifying the environmental impacts of consumption and production, which pinpoints 'hotspots' such as the agriculture and fossil fuel value chains as where change is needed and unexpected consequences of current consumption and production patterns need to be addressed.

4.6 *Life Cycle Initiative Networks—Growing in Numbers and Expertise*

The capability development efforts of the Life Cycle Initiative aim to empower individuals and societies with the necessary skills and competences to move our societies toward more sustainable production and consumption patterns. Together with our partners we are building networks, sharing knowledge, data, experiences and best practices, and implementing projects to foster a new generation of citizens who integrate life cycle thinking in their personal and professional decisions. Target audiences of our capability development activities include: (1) scientists, (2) business, (3) governments and the (4) civil society.

An overview of current local and national networks of life cycle practitioners and stakeholders worldwide is given in Fig. 4.3. The Life Cycle Initiative has been working over the past decade to broaden the base of practitioners and stakeholders working with life cycle approaches around the world, and to establish links between those working in the same countries and regions. The map indicates the networks with which the Life Cycle Initiative has regular contact for discussions, feedback, and assistance in the preparation and rollout of materials.

4.6.1 The International Life Cycle Network

One of the greatest successes of the Life Cycle Initiative's 10 years of existence has been the establishment of a global network of life cycle practitioners

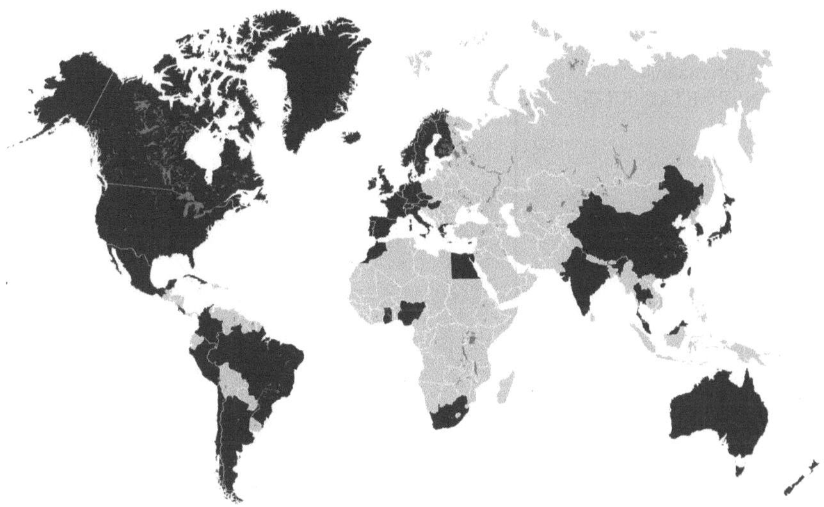

Fig. 4.3 Overview of current national life cycle networks worldwide. (UNEP/SETAC 2012a)

(http://www.lifecycleinitiative.org/networks/). Currently with over 2,000 registered participants located in about 50 international, national and regional networks in Asia, Latin America, Europe, the USA, Africa, the Middle East and Central Asia, the network is continually growing (Fig. 4.4). The successful establishment of the network has been based on a good understanding of the needs of the users of life cycle tools attained through surveys and stakeholder consultations. Their major activities relate to knowledge sharing and communication, support of case studies, and development of life cycle inventories and impact assessment methods. The network is now a self-sustaining entity and supports the Life Cycle Initiative's work plan by providing the needed body of experts to complete peer reviews, as well as being a source of input and consensus on new tools and guidance. This first comprehensive online mapping can serve as a basis for enhancing cooperation and coordination among the LCA networks worldwide (Bjørn et al. 2013).

While the roots of life cycle thinking and expertise is found in Europe, jointly with North America and Japan, the Life Cycle Initiative has worked with partner organizations to launch regional life cycle networks in Africa and Latin America, as well as national networks in China, Argentina and Colombia. New networks are also being built in India, Russia and in some African countries.

4.6.2 Life Cycle Jobs are Green Jobs

Life cycle practitioners are in demand in a green economy. This has already been recognized by some countries, as described in the ILO submission to the RIO+20 Summit on how occupations change as an economy goes green. The report makes

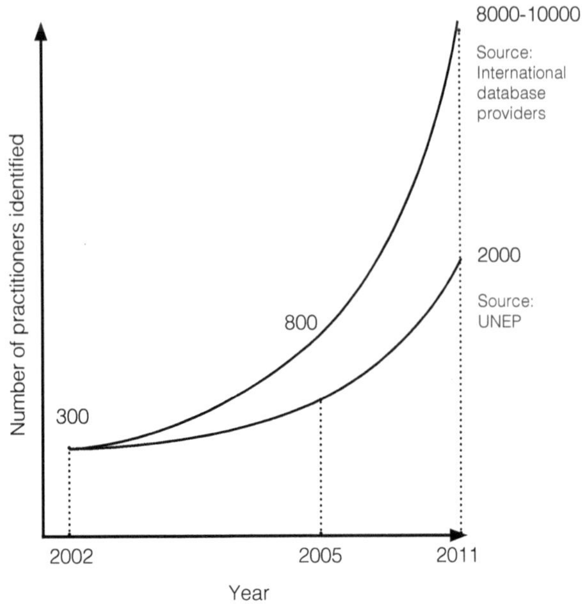

Fig. 4.4 Growth of registered participants in the life cycle initiative's international network. (UNEP/SETAC 2012a)

several references to need for life cycle assessment skills, for example, in the checklist of green skills identified by the UK government. In a further example, the Republic of Korea is noted to have created new 'sector skills councils', one of which provides training on sustainability assessment.

4.6.2 Accomplishments in Phases 1 and 2

Phases 1 and 2 activities prioritized the strengthening of regional and international scientific networks worldwide. The network database now lists more than six times the number of entries since it debuted. Projects to expand the network and develop life cycle capability included dozens of seminars, workshops and sessions at international conferences and meetings, particularly focusing in non-OECD countries.

Another indication of the number of international life cycle stakeholders is the number of hits per month on the Life Cycle Initiative's portal with online tools, which total more than 15,000. The Global Guidance for LCA Databases, the Social LCA and LCM training materials are the most downloaded documents. Materials are broadly used in developing economies and businesses around the world for internal training.

The 'UNEP/SETAC LCA Award for non-OECD countries' acknowledges the work from academics and private companies in developing and emerging economies who have started visionary and innovative projects based on the life cycle approach. The first three editions of the prize were given in 2006, 2008 and 2010. In 2010, thirteen projects received a 1-year license of a life cycle software.

4.7 *Communicating Life Cycle Information—The Right Story for Every Audience*

Life cycle assessment consists of the identification and assessment of impacts along a product's value chain and then, communication of the result in a useful way so that the information can be used for decision-making. The main goal of communicating life cycle information in the transition to a green economy then is to induce change toward more sustainable decision-making from all stakeholders on process, products and organizations.

The main providers of Life Cycle information, also called Environmental Product Information, are industry and businesses, i.e. the supply side. The latter are motivated by a series of driving forces, which depend on the target audience and which include the communication of Environmental Product Information:

- Final private consumers, in order to get competitive advantage in emerging or new green markets;
- Business clients, either because requested to (this is especially the case of SMEs in the supply chain), or to compete in the business-to-business market arena;
- Societal and other stakeholders, to respond to the external pressure from environmental NGOs and consumer associations, and to convey a more holistic life cycle picture of products and services, in order to induce an appropriate use and disposal of products;
- Financial stakeholders, who are increasingly attentive to the sustainable dimensions of organizations and products;
- Public administrations, in order to apply to Green Public Procurement programmes and/or to obtain tax incentives, whenever applicable;
- Policy makers, providing credible life cycle information and reference data to support them in better-informed policy decisions and to prevent a misuse of life cycle approach and simplistic green claims, which might be highly misleading.

Life cycle information can be communicated in many ways, with varying levels of detail, considering various parts of the life cycle, different pillars of sustainable development (i.e., environmental/economic/social), and with varying levels of external verification. Some address a broad range of indicators, and some focus on one or two indicators. Some are based on full life cycle assessments and some on life cycle thinking. To bring some structure into this wide range of communication possibilities, ISO has put in place the ISO 14020 series of environmental standards. Key differences exist between communication in the form of corporate reporting, from business to business and from business to consumers as well as to motivate sustainable consumption behaviour.

In the context of the Life Cycle Initiative the brochure 'Why take a life cycle approach?' and the document 'Communication of Life Cycle Information in the Building and Energy Sectors' were published or contributed to as part of the Life Cycle Initiative's work programme on communicating life cycle information, including general awareness raising/educational materials, as well as publications for practitioners.

5 The Future of Life Cycle Thinking and Phase 3 of the Life Cycle Initiative

5.1 Consultation Process

In 2011 and 2012, the Life Cycle Initiative began an extensive consultation process involving an outreach survey, focus group discussion and a meeting with experts with the results to be used as a starting point for its Phase 3 strategy development process. More than 200 LCA practitioners and partners provided very valuable feedback, which is summarized in the following replies to the questions raised:

- What is limiting more implementation of life cycle approaches in your country or industry?
  - Easy access to reliable data,
  - Lack of business drivers,
  - Lack of awareness/understanding,
  - Cost issues,
  - Unclear relationship among a number of related tools such as LCA, carbon footprinting and water footprinting and concepts such as LCM and Corporate Social Responsibility,
  - Lack of harmonisation between methods,
  - Difficult to engage small and medium sized enterprises.
- As an LCA professional, what would be your aspirations for 2017?
  - LCA has been introduced into education programmes: e.g., life cycle thinking in schools and LCA courses in higher education;
  - LCA is daily practice in business and industry: e.g., at least all big companies use LCA results;
  - Life cycle thinking widely accepted as a basis for decision making: e.g., LCA results are used as key information for decision-making by government, industry, and by consumers;
  - Capability on LCA has been enhanced in non-OECD countries: e.g. a market for life cycle expertise has been established in all relevant emerging economies;
  - Life cycle information on products is available in one way or the other to consumers: e.g. consumers can get information on the environmental footprint of different product groups from internet.
- What are your expectations for the Life Cycle Initiative during the next 3–5 years?
  - Ensure consistency in the way data sets are developed around the world and support access to better and more data,
  - Non-OECD engagement to achieve a minimum standard of LCA usage worldwide,
  - Harmonisation/consensus building on methodologies,
  - Continued methodological development,

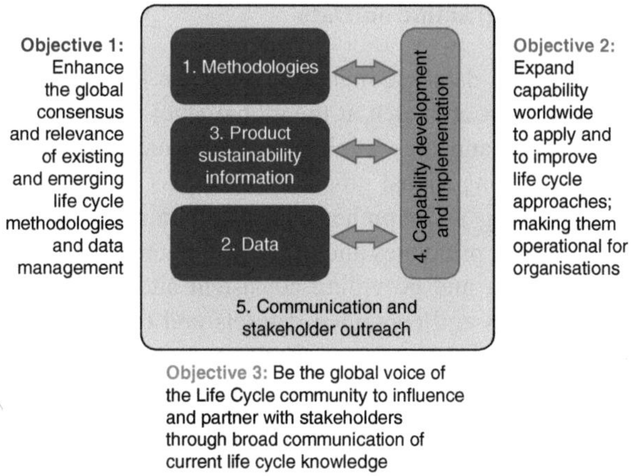

Fig. 4.5 Programme areas to achieve the objectives of phase 3

- Training of practitioners,
- Communication/awareness raising/lobbying,
- Education of key decision-makers in industry and government,
- Partnering, Case study development, Dissemination, Engaging stakeholders.

5.2 New Strategic Approach and Programmes

Building on the results of the expert consultation and on an assessment of strengths, weaknesses, opportunities and threads, the Life Cycle Initiative has developed the new vision of 'a world where life cycle approaches are mainstreamed' with a supporting strategic approach. The Life Cycle Initiative started Phase 3 in 2012 with a mission to "enable the global use of credible life cycle knowledge for more sustainable societies". Its overarching goal is to "facilitate the generation and uptake of science-based life cycle approaches and information for products and organization by business, government and civil society practice worldwide as a basis for sustainable consumption and production".

As illustrated in Fig. 4.5, the overall objective is backed by three specific objectives that are related to five programme areas for implementation, within which flagship projects have been identified. Although flagships have been recognised other relevant activities such as the consolidation of USEtox, the further development of characterization factors for water and land use in LCA and the dissemination of the LCM capability maturity model will continue as renewed projects in Phase 3. Also ideas for new 'normal' projects have been retained such as the development of a code of conduct for LCA professionals, the provision of guidance on the use of input-output/hybrid data and the set up of multistakeholder roundtables to address relevant challenges.

The following seven flagship projects have been prioritized and articulated within the five programmes.

5.2.1 Programme on Data

A. Data and database management—The focus of this flagship is on promoting a consistent approach at the global level to the creation and maintenance of LCI datasets and the development and management of LCA databases based on the existing global guidance principles for LCA databases. This is done by (a) producing a comprehensive set of training material in support of the Shonan guidance principles and using this material in various countries and regions, (b) preparing and providing consistent and accessible informational resources on databases and contained datasets and (c) establishing and supporting regional and global networks of database developers and managers (in close cooperation with the activities on capability development, see flagship G). The flagship includes also technical assistance activities if requested as well as the expansion and enhancement of the LCI database registry (in close cooperation with the new life cycle platform, see flagship H).

5.2.2 Programme on Methodologies

B. Life Cycle Sustainability Approaches and Organisations– Some of the tools developed during Phase 2 have not yet reached full maturity in order to be used on a broad scale. Efforts of this flagship will focus on further development and testing of S-LCA and LCSA in particular and also on challenges with regards to LCA for organisations.

C. Environmental life cycle impact assessment indicators—The objective of this flagship project is to run a global process aiming at global guidance and consensus building on a limited number of life cycle impact category indicators developed within a consistent framework and to identify the related research agenda. The deliverable would be one or more global guidance publications with a supporting web system that includes a limited number of six to ten LCA based environmental impact category indicators and the characterisation factors (for various regions). It may also include guidance how to best establish a particular regional impact category indicator in case global consensus on characterisation factors cannot be achieved or makes no sense.

5.2.3 Programme on Product Sustainability Information

D. Product environmental information meta 'specification'—Multiple efforts are on-going in different parts of the world to develop an authoritative identification of the hotspots and most relevant environmental impacts and resource uses for a certain product categories or group. This flagship project seeks to provide guidance on the broader considerations that should be taken into account, and the key principles that could be applied for different product sustainability information systems to allow more informed decision making by purchasers.

E. Knowledge mining guidance—There are hundreds of existing LCA studies that, taken together, represent a significant base of knowledge that can be tapped into. The aim of this flagship is to provide a methodology for mining knowledge from these LCA studies, using one or more pilot studies to demonstrate the value of this process (starting with food packaging sector). Can we use the review of existing studies to identify critical messages or themes that might inform policy makers in government and industry? What institutional buyers or consumers can learn from these studies with regard to how they have translated the technical and scientific information into a language that can be understood?

5.2.4 Programme on Capacity Building and Implementation

F. Global capability development—This flagship project has the aim to strengthen and consolidate the life cycle work in the regions, including documentation of local consultants and databases available. Focal points at Governmental offices (including national statistic offices for data management aspects) and chambers of commerce will be identified and linked to the national networks. Some deliverables identified for this flagship include:

- Establishing a baseline on the level of life cycle thinking worldwide, assessing the current capabilities on life cycle approaches in non-OECD countries, with updates planned for every 3 years to trace the evolution;
- Life cycle tools (i.e. on life cycle management, life cycle based footprinting indicators and ecodesign) spread across the emerging and rapidly growing economies via the Life Cycle Initiative's or local platforms;
- South-South (e.g. in Latin America) cooperation for increased implementation and North-South cooperation for methodologies' enhancement, data generation and exchange;
- Life cycle experts and practitioners network established in each region of the world;
- Online tools, if possible, translated into several languages including Chinese, French, Portuguese, and Spanish, in addition to English.

5.2.5 Programme on Communication and Stakeholder Outreach

G. Communicating life cycle knowledge—The main deliverable for this flagship is an improved, informative, interactive and educational virtual platform with a communication strategy based on social media tools. The new life cycle platform at http://www.lifecycleinitiative.org will contribute to the life cycle awareness by sharing relevant knowledge and data. Information to be shared can be extracted from significant life cycle studies, key business websites presenting clear benefits (and challenges) from implementing life cycle approaches, the life cycle inventory database registry, appropriate websites from similar and related initiatives sharing online tools, and so on. The life cycle platform will also help identifying the most relevant life cycle trends. A next step is the stakeholder outreach by various means, including the above-mentioned multi-stakeholder roundtables. New features of the website include Facebook, twitter and LinkedIn applications.

5.3 Setting up the Baseline for Phase 3 of the UNEP/SETAC Life Cycle Initiative—Monitoring Progress by Key Indicators

The Phase 3 elements of the strategic document and the action plan developed for flagship projects' implementation will guide the work of the UNEP/SETAC Life Cycle Initiative from 2012 through 2017. In order to monitor progress of their implementation, outcome indicators are essential. UNEP and SETAC suggested using them and the International Life Cycle Board and the life cycle community supported this proposal. The outcome indicators are expected to measure the quantity and quality of the results achieved by the activities deployed by the UNEP/SETAC Life Cycle Initiative.

Essentially the indicators reflect how the business, local and national Governments, academia and civil society are taking up the recommendations and deliverables produced by the UNEP/SETAC Life Cycle Initiative. Examples of outcome indicators include number of life cycle network and associations worldwide being in contact with the initiative or number of individuals or organisations using the USEtox model.

6 Conclusions and Perspectives

The key achievements of the UNEP/SETAC Life Cycle's Phase 1 and 2 activities are summarised below.

- Global life cycle community—the activities of the Life Cycle Initiative such as the various editions of the LCA Award for non-OECD countries have been crucial for the dissemination of LCA worldwide and the creation of a global community, including the set up of national and regional life cycle networks in different parts of the world.
- The Life Cycle Inventory Database Registry—this is a global repository for finding and offering LCA related datasets of high quality. The project motivated the European Commission to develop their ILCD Data Network.
- Life Cycle Impact Assessment Midpoint-damage framework—this framework links a product's environmental interventions or impacts to the ultimate effects on human health, ecosystem quality and resource depletion.

- USEtox—an environmental model for characterization of human and ecotoxic impacts in Life Cycle Impact Assessment and for comparative assessment and ranking of chemicals according to their inherent hazard characteristics.
- Life Cycle Management guide—the guide provides a coherent methodology for implementing life cycle approaches and activities with the goal of continual performance improvement. Life cycle thinking is made operational through Life Cycle Management. LCM is a management approach that puts the tools and methodologies in the life cycle thinking basket into practice.
- Life Cycle Management Capability maturity framework—the framework shifts the focus from driving performance on prescriptive sustainability metrics to building the capacity of organizations in a supply chain to identify and manage social and environmental issues in a manner that is tailored to their business strategy. It helps the supplier to identify where and how to start and continue their journey towards sustainability.
- Guidelines on Social LCA—these guidelines set out the key elements, indicators, and limitations for assessing the positive and negative social impacts of a product over its life cycle, with consideration of human rights, working conditions, health and safety, among others. This is a final key element that will enable a full triple-bottom-line approach to product sustainability assessment.
- Life Cycle Sustainability Assessment—this framework brings together the two established product assessment tools of environmental LCA and life cycle costing with the newly developed social LCA to establish the process for a triple-bottom-line assessment of a product's life cycle impacts.
- Global Guidance Principles for Life Cycle Assessment databases: these principles give guidance for proper gathering and management of data, which enable better, more reliable life cycle assessment results and improve their use for decision-making.
- Work on methodologies with regard to carbon footprinting to bridge the standard developments with regard to LCA done by the WBCSD/WRI Greenhouse Gas Protocol and ISO as well as on water and land use in LCA have ensured that the available LCA knowledge in the life cycle community is considered for carbon footprinting and water footprinting standards and that new widely accepted characterisation factors are about to be developed for water and land use impacts on biodiversity and ecosystem services.

Building on the achievements from Phases 1 and 2 and in particular the results of a stakeholder consultation process in 2011 and 2012, the vision for Phase 3 was coined as 'a world where life cycle approaches are mainstreamed'. The journey of 10-years Life Cycle Initiative is illustrated in Fig. 4.6. Activities in Phase 3 will focus on creating the enabling conditions to (a) enhance the global consensus and relevance of existing and emerging life cycle methodologies and data management; (b) expand capabilities worldwide and make life cycle approaches operational for organizations; and (c) communicate current life cycle knowledge to influence and partner with stakeholders. In consultation with the International Life Cycle Board, seven flagship projects have been defined in the areas of methodologies, data, product sustainability

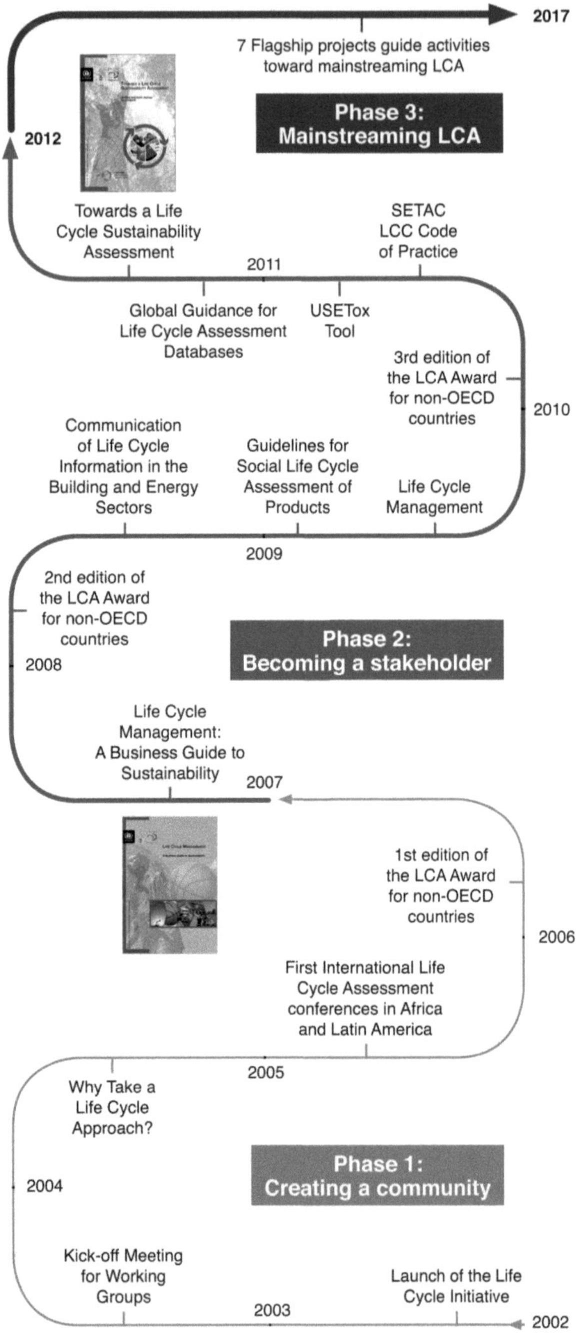

Fig. 4.6 The key achievements of the UNEP/SETAC life cycle in its journey of 10 years. (UNEP/SETAC 2012a)

information, capability development and implementation, and communication and stakeholder outreach. They are expected to be implemented jointly with a number of other projects. Progress made in Phase 3 will be monitored every two to 3 years by key indicators and compared to a baseline survey carried out in 2012.

Acknowledgements Acknowledgements to the ILCB members and donors of the UNEP/SETAC Life Cycle Initiative for providing the strategic advice and financial resources needed. See http://www.lifecycleinitiative.org for full list of current partners and further information on the initiative.

Moreover, the authors would like to thank Bruce Vigon from SETAC and Jim Fava as PMO co-chair for their support in moving the Life Cycle Initiative into Phase 3. Finally, the authors would like to recognize the work of Thad Mermer who helped the secretariat of the Life Cycle Initiative to develop the key messages.

Disclaimer The designations employed and the presentation of the material in this publication does not imply the expression of any opinion whatsoever on the part of the UNEP/SETAC Life Cycle Initiative concerning the legal status of any country, territory, city or area or of its authorities, or concerning delimitation of its frontiers or boundaries. Moreover, the views expressed do not necessarily represent the decision or the stated policy of UNEP or SETAC, nor does citing of trade names or commercial processes constitute endorsement.

Appendix—Glossary

10YFP	10-Year Framework of Programmes on sustainable consumption and production adopted during the last so-called Rio + 20 World Summit on sustainable development in June 2012
Apeldoorn declaration	Declaration by a group of specialists in LCA and Risk Assessment on practices and complications of life cycle impact assessment methodologies for non-ferrous metals
CILCA	International Conference on life cycle assessment in Latin America
CML	Institute of Environmental Sciences, an institute of the Faculty of Science of the Leiden University
E-LCA	Environmental Life Cycle Assessment
EPFL	Ecole Polytechnique Fédérale de Lausanne
ICMM	International Council on Mining and Metals
ISO	International Organization for Standardization
ILCB	International Life Cycle Initiative Board
ILCD	International Reference Life Cycle Data System developed by the European Commission
ILCP	International Life Cycle Initiative Panel
IRP	International Resource Panel
Life Cycle Initiative	UNEP/SETAC Life Cycle Initiative
LCC	Life Cycle Costing
LCI	Life Cycle Inventory
LCIA	Life Cycle Impact Assessment

LCM	Life Cycle Management
LCSA	Life Cycle Sustainability Assessment
Marrakech Process	A global process as called for by the World Summit on sustainable development's Johannesburg plan of action to support the elaboration of a 10-Year framework of programmes on sustainable consumption and production
OECD	Organisation for Economic Cooperation and Development
PMO	Project Management Office
SCP	Sustainable consumption and production
SETAC	Society of Environmental Toxicology and Chemistry
Shonan Guidance Principles	Global guidance principles for life cycle assessment databases
S-LCA	Social life cycle assessment
TF	Task force
UNEP	United Nations Environment Programme
USEtox	Environmental model for characterization of human and ecotoxic impacts in life cycle impact assessment and for comparative assessment and ranking of chemicals according to their inherent hazard characteristics developed by a team of researchers from the Phase 1 Task force on toxic impacts under the UNEP-SETAC life cycle initiative
WAIG	Work Area Interest Group
WBCSD	World Business Council for Sustainable Development
WRI	World Resource Institute

References

Aloisi de Larderel J, Fava J (2003) Int J LCA—associated journal of the UNEP/SETAC life cycle initiative. Int J Life Cycle Assess 8(3):117

Bayart J-B, Bulle C, Deschênes L, Margni M, Pfister S, Vince F, Koehler A (2010) A framework for assessing off-stream freshwater use in LCA. Int J Life Cycle Assess 15(5):439–453

Benoît-Norris C, Vickery-Niederman G, Valdivia S, Franze J, Traverso M, Ciroth A, Mazijn B (2011) Introducing the UNEP/SETAC methodological sheets for subcategories of social LCA. Int J Life Cycle Assess 16(7):682–690

Bjørn A, Owsianiak M, Laurent A, Molin C, Westh TB, Hauschild MZ (2013) Mapping and characterization of LCA networks. Int J Life Cycle Assess 18:812–827

Coleman T, Masoni P, Dryer A, McDougall F (2003) International expert group on life cycle assessment for integrated waste management. Int J Life Cycle Assess 8(3):175–178

Curran MA (2006) Report on activity of task force 1: data registry—global life cycle inventory data resources. Int J Life Cycle Assess 11(4):284–289

Datta A, Strothmann S, Valdivia S, Vigon B (2012) First Indian life cycle assessment and management conference 2012. Int J Life Cycle Assess 17(7):944–945

Diamond ML, Gandhi N, Adams WJ, Atherton J, Bhavsar SP, Bulle C, Campbell PGC, Dubreuil A, Fairbrother A, Farley K, Green A, Guinee J, Hauschild MZ, Huijbregts MAJ, Humbert S, Jensen KS, Jolliet O, Margni M, McGeer JC, Peijnenburg WJGM, Rosenbaum R, van de Meent D, Vijver MG (2010) The clearwater consensus: the estimation of metal hazard in fresh water. Int J Life Cycle Assess 15(2):143–147

Fava J, Udo de Haes HA, Sonnemann G, Dubreuil A, Gloria T (2006) Key observations arising from papers on sustainable production, use and recycling of natural resources. Int J Life Cycle Assess 11(1):75–76

Finkbeiner M (2009) Carbon footprinting—opportunities and threats. Int J Cycle Assess 14(2):91–94
Hauschild MZ, Jolliet O, Huijbregts MAJ (2011) Editorial. LCIA of impacts on human health and ecosystems (USEtox)—a bright future for addressing chemical emissions in life cycle assessment. Int J Life Cycle Assess 16(8):697–700
Hellweg S, Demou E, Bruzzi R, Meijer A, Rosenbaum R, Huijbregts MAJ, McKone TE (2009) Integrating human indoor air pollutant exposure within life cycle impact assessment. Environ Sci Technol 43(6):1670–1679
ISO 14040 (1997) Environmental management—life cycle assessment—principles and framework. Geneva
ISO 14046.CD.1 (2012) Water footprint—requirements and guidelines. Geneva
ISO/TS 14067 (2013) Carbon footprint of products—requirements and guidelines for quantification and communication. Geneva
Jolliet O, Müller-Wenk R, Bare J, Brent A, Goedkoop M, Heijungs R, Itsubo N, Peña C, Pennington D, Potting J, Rebitzer G, Stewart M, Udo de Haes HA, Weidema B (2004) The LCIA midpoint-damage framework of the UNEP/SETAC life cycle initiative. Int J Life Cycle Assess 9(6):394–404
Jolliet O, Dubreuil A, Gloria T, Hauschild M (2005) Progresses in life cycle impact assessment within the UNEP/SETAC life cycle initiative. Int J Life Cycle Assess 10(6):447–448
Koellner T, de Baan L, Beck T, Brandao M, Civit B, Goedkoop M, Margni M, Milà i Canals L, Müller-Wenk R, Weidema B, Wittstock B (2013) Principles for life cycle inventories of land use on a global scale. Int J Cycle Assess 18:1203–1215
Lundie S, Huppes G, Ciroth A (2007) Inventory methods in LCA: towards improved methodological consistency—Report for UNEP-SETAC life cycle initiative life cycle inventory (LCI) programme task force 3 on methodological consistency, ISBN 3836486210
Malmö Ministerial Declaration (2000) Global Ministerial Environment Forum—Sixth special session of the governing council of the United Nations Environment Programme. Nairobi
Milà i Canals L, Bauer C, Depestele J, Dubreuil A, Freiermuth Knuchel R, Gaillard G, Michelsen O, Müller-Wenk R (2007) Key elements in a framework for land use impact assessment within LCA. Int J Cycle Assess 12(1):5–15
Rosenbaum RK, Bachmann TM, Gold LS, Huijbregts MAJ, Jolliet O, Juraske R, Koehler A, Larsen HF, MacLeod M, Margni M, McKone TE, Payet J, Schuhmacher M, Van de Meent D, Hauschild MZ (2008) USEtox—the UNEP-SETAC toxicity model: recommended characterisation factors for human toxicity and freshwater ecotoxicity in life cycle impact assessment. Int J Life Cycle Assess 13(7):532–546
SETAC (2005) Life-cycle assessment of metals: issues and research directions. Pensacola, 307 pp
SETAC (2011) Environmental life cycle costing: a code of practice. Pensacola, 98 pp
Sonnemann G (2003) Start of the next phase: task forces and potential future crosscutting activities. Int J Life Cycle Assess 8(6):323
Sonnemann G (2004a) Crosscutting issues to be explored by the UNEP/SETAC life cycle initiative. Int J Life Cycle Assess 9(1):67
Sonnemann G (2004b) Strengthening capacity building through regional networks. Int J Life Cycle Assess 9(5):334
Sonnemann G, Valdivia S (2007) First phase 2 activities of the UNEP/SETAC life cycle initiative. Int J Life Cycle Assess 12(7):544–545
Sonnemann G, Solgaard A, Saur K, Udo de Haes HA, Christiansen K, Jensen AA (2001) Life cycle management: UNEP-workshop—sharing experiences on LCM. Int J Life Cycle Assess 6(6):325–333
Sonnemann G, Quiros A, Musmanni S, Fava J, Schenk R, Ugaya C (2005) International conference on life cycle assessment 2005 (CILCA 2005), 25–28 April 2005, San Jose, Costa Rica. Int J Life Cycle Assess 10(2):163
Sonnemann G, Vigon B, Broadbent C, Curran MA, Finkbeiner M, Frischknecht R, Inaba A, Schanssema A, Stevenson M, Ugaya C, Wang H, Wolf M-A, Valdivia S (2011) Process on 'global guidance for LCA databases'. Int J Life Cycle Assess 16(1):95–97
Swarr T (2011) A capability framework for managing social and environmental concerns. Int J Cycle Assess 16(7):593–595

Töpfer K (2002) Editorial for Int J LCA on the launch of the UNEP-SETAC life cycle initiative. Int J Life Cycle Assess 7(4):191
Udo de Haes HA, Jolliet O, Norris G, Saur K (2002) UNEP/SETAC life cycle initiative—background, aims and scope. Int J Life Cycle Assess 7(4):192–195
UNEP (2003) Evaluation of environmental impacts in life cycle assessment. Paris, 96 pp
UNEP/SETAC (2004) Why take a life cycle approach? Paris, 24 pp
UNEP/SETAC (2007) Life cycle management—a business guide to sustainability. Paris, 52 pp
UNEP/SETAC (2008) Communication of life cycle information in the buidling and energy sector. Paris, 176 pp
UNEP/SETAC (2009a) Guidelines for social life cycle assessment of products. Paris, 103 pp
UNEP/SETAC (2009b) Life cycle management—how business uses it to decrease footprint, create opportunities and make value chains more sustainable. Paris, 48 pp
UNEP/SETAC (2011a) Global guidance principles for life cycle assessment databases—a basis for greener processes and products. Paris, 158 pp
UNEP/SETAC (2011b) Towards a life cycle sustainability assessment—making informed choices on products. Paris, 65 pp
UNEP/SETAC (2012a) Greening the economy through life cycle thinking—10 years of the UNEP/SETAC life cycle initiative. Paris, 60 pp
UNEP/SETAC (2012b) Phase 2: 2007–2012. http://www.lcinitiative.unep.fr. Accessed 30 Oct 2012
WBCSD/WRI (2011) Greenhouse gas protocol product standard. Geneva

Chapter 5
Life Cycle Assessment as Reflected by the International Journal of Life Cycle Assessment

Almut B. Heinrich

Abstract Six 'milestones' in the life of the journal (until 2013) are identified in this chapter and the accompanying literature is discussed:

1. Institute for Scientific Information (ISI)—impact factor
2. Online publications
3. National societies and other organisations and networks
4. New topics and subject areas
5. Special issues and supplements
6. ISO standardisation of LCA

These 'milestones' have significantly impacted the development of Int J Life Cycle Assess and, thereby, that of the field of LCA.

The sections of this chapter demonstrate that Int J Life Cycle Assess

- has been a truly international journal from the beginning,
- addresses the global LCA community,
- offers a unique spectrum of LCA-related information,
- applies to scientists, practitioners, consultants, governments and administration,
- responds to the growing awareness that life cycle-based assessment methods are unique achieving sustainability,
- the field of LCA and Int J Life Cycle Assess have interacted and mutually benefited.

Keywords CHAINET · Digital object identifier · Impact factor · Institute for Scientific Information (ISI) · Int J Life Cycle Assess · ISO standardisation of LCA · ISOLP · LCA · LCANET · Life cycle assessment · Life cycle assessment in Australia · Life cycle assessment in India · Life cycle assessment in Japan · Life cycle assessment in Korea · Life cycle assessment in New Zealand · Life Cycle Costing (LCC) · Life Cycle Management (LCM) · National societies · Online publications · SETAC · Social Life Cycle Assessment (SLCA) · Special issues · SPOLD · Subject areas · Supplements · Swiss Discussion Forum on Life Cycle Assessment · UNEP/SETAC Life Cycle Initiative

---

A. B. Heinrich (✉)
Steinhofweg 63, 69123, Heidelberg, Germany
e-mail: abh.scientificjournals@googlemail.com

W. Klöpffer (ed.), *Background and Future Prospects in Life Cycle Assessment*,
LCA Compendium – The Complete World of Life Cycle Assessment,
DOI 10.1007/978-94-017-8697-3_5, © Springer Science+Business Media Dordrecht 2014

1 Introduction

The following two statements from The International Journal of Life Cycle Assessment (Int J Life Cycle Assess) are programmatic:

- "This Journal … is the first to be devoted entirely to the science and practice of LCA. It is conceived as an international scientific journal … " (Klöpffer 1996, editorial).
- " … we offer a unique spectrum of LCA-related information, indispensable for the whole LCA community" (Klöpffer and Heinrich 1999, editorial).

This chapter demonstrates the trueness of these statements.

"Hitherto, publication in the field of LCA has been restricted to Workshop Reports, the LCA-Newsletters, and much 'gray literature'. Very recently, Environmental Toxicology & Chemistry, published by SETAC, provided access to LCA-type articles. Several journals specialised in environmental sciences, packaging, surfactants, etc. sporadically publish papers related to LCA" (Klöpffer 1996, editorial).

With the launch of Int J Life Cycle Assess in December 1995, publishing in the field of LCA (Life Cycle Assessment) changed completely. The Journal, conceived as an international scientific journal, was the first to be devoted entirely to the science and practice of LCA. It continues to be the only regularly published journal dedicated to LCA.

The establishment of Int J Life Cycle Assess was fully acknowledged and confirmed on the occasion of Walter Klöpffer's birthday in 2008 by members of the Editorial Board (Hunkeler et al. 2008):

> The LifeCycle: Vast amount of aspects—many different interpretations—few global players—one relevant journal! Martin Baitz
>
> Walter and his colleagues had already been active in LCA. Since then Walter has taken a visionary position in striking new ground with the Journal of LCA—at that time few thought it would succeed—not enough interest in this new but growing tool. He was right. His leadership with the journal, willingness to partner with groups to advance LCA globally, breaking ground on the LCA study peer review process, and improving the quality of the LCA papers in the journal have all been some of the success factors that has established LCA as a solid tool in business and governments to improve our product and packaging systems. James Fava
>
> During the early 90's it was quite difficult to get LCA papers published in scientific journals. This hampered the development of LCA and the acceptance of LCA results. Starting The International Journal of LCA was a key step in the development of LCA as a tool. This has been immensely important for development of LCA as a scientific discipline and the acceptance of LCA as a tool. Göran Finnveden
>
> When the LCA community was fully working within the SETAC organisation, Walter came up suddenly with the surprising idea of establishing an international journal on LCA, fully independent from our scientific home. Surely we found this journal a very good idea, there was a real need for it; but why in this way? Would it work? And now we can say, yes, look how good this idea was, precisely in the way he proposed. Helias A. Udo de Haes

The following sections show how the global LCA community is addressed by Int J Life Cycle Assess, both geographically and topically.

Section 2 identifies 'milestones' in Int J Life Cycle Assess.

Sections 3 and 4 discuss two 'external' elements that opened the way to the global and topical adoption and proliferation of the journal and, therewith, to the field of LCA. These are the impact factor and the digital object identifier; the latter was the pre-requisite for the online editions.

Section 5 describes the national societies and other organisations, networks and initiatives which continue to contribute considerably to the success of the journal.

Section 6 deals with the great variety of new topics and subject areas which mirror the development of LCA in the journal.

Section 7 covers the many special issues and supplements mapping the acceptance of the journal and its geographical and topical coverage.

Section 8 compiles the articles on the ISO standardisation process of LCA in the journal.

2 Milestones in Int J Life Cycle Assess

The following 'milestones' in the life of the journal impacted the development of Int J Life Cycle Assess[1], and represent the themes of this chapter.

1. Institute for Scientific Information (ISI)—impact factor
   On September 25, 2001, ISI (Institute for Scientific Information) accepted Int J Life Cycle Assess for coverage in Science Citation Index Expanded, beginning with issue no. 1 of vol. 6, 2001.
   In addition, Int J Life Cycle Assess was included in ISI Web of Science, ISI Alerting Services, and Current Contents/Agriculture, Biology, and Environmental Sciences.
   This was the prerequisite for calculating an Impact Factor.
2. Online publications
   In issue no. 6 of vol. 4, 1999, it was announced that, by means of the DOI (digital object identifier), articles can be published shortly after their acceptance, which means weeks or months before they can appear in the printed journal and even before the issue number and the true page numbers have been determined.
3. National societies and other organisations and networks
   The national societies as well as a number of other organisations and networks reflect the proliferation of the journal and simultaneously the global adoption of the field of LCA.

By 2000, Int J Life Cycle Assess was the Official Organ of four societies: JLCA (LCA Society of Japan), ISLCA (Indian Society for LCA), KSLCA (Korean Society

[1] The journal was founded and published, from 1996 to 2007, by ecomed publishers, Landsberg/Lech, Bavaria (Germany). Then the abbreviation of the journal was 'Int. J. LCA'. When the journal was transferred to Springer-Verlag in 2008 (Roos 2007), the abbreviation changed to 'Int J Life Cycle Assess'.

for LCA), and ALCAS (Australian LCA Society). LCANZ (Life Cycle Association of New Zealand) joined in 2005/2010. Of these four societies, JLCA has been the most active.

In 2003, Int J Life Cycle Assess became the 'Associated Journal of UNEP/SETAC Life Cycle Initiative'.

4. Topics and subject areas
5. Special issues and supplements
6. ISO standardisation of LCA

3 Institute for Scientific Information (ISI)—Impact Factor

Since 2001, Int J Life Cycle Assess has been listed in the Science Citation Index (the Science Citation Index Expanded, to be exact).

There are two striking papers on the 'philosophy' of the impact factor as it relates to Int J Life Cycle Assess:

1. 'Int J LCA Could Have Received Better Acknowledgement' by Henrikke Baumann (Baumann 2002)
2. 'Publishing Scientific Articles with Special Reference to LCA and Related Topics' by Walter Klöpffer (Klöpffer 2007, p. 73)

Henrikke Baumann criticised that ISI did not allocate the journal to the Science Citation Index but rather to the Science Citation Index Expanded. She argued: "Inclusion in the SCIExp [Science Citation Index Expanded] is a good acknowledgement, but inclusion in the SCI (Science Citation Index) is better. The Institute for Scientific Information (ISI) covers 8,000 journals annually to produce the SCI, the SCIExp [now SciSearch] and other related information products (ISI 2001). The SCIExp covers approx. 5,800 leading scientific and technical journals, while the SCI includes only 3,800 journals which are considered to be the most influential, i.e. the SCI is a subset of the SCIExp. The point is, Int J LCA could have done better. It could have been included in the SCI, if only ISI had got access to more profound citation statistics" (Baumann 2002).

Admittedly, the people at ISI had to first review the journal in terms of its regular appearance, the appropriate frequency, as well as the scientific approach, contents and setting. However, it took them five years (1996–2000) to accept the journal for coverage. The reason may be that, being the first environmental life cycle assessment journal (Klöpffer 2007), Int J Life Cycle Assess did not belong to the 'classical' scientific categories within ISI. Therefore, it may have been eyed with scepticism. This may also be the reason for the deviation via the Science Citation Index Expanded. This deviation has not yet been changed, although the 'access to more profound citation statistics' should have been given in the meantime.

An actual overview of all Abstracting and Indexing services covering Int J Life Cycle Assess can be found at:

http://www.springer.com/environment/journal/11367

4 Online Publications

The ecomed publishers introduced the DOI in 1998. With this, the suffix could be created which reflected the journal, the year of publication, the month of publication and the running number. It followed a very creative phase of building the directories of the online editions, jointly with the most capable and visionary webmaster Rainer Schwandt.

"With Online-First, articles can be published shortly after their acceptance by reviewers, authors, and editor, which means weeks or months before they can appear in [the printed] Int. J. LCA and even before the issue number and the true page numbers have been determined. … Publication date of the article is the Online-Publication Date which is indicated in both Online-First and the printed article in Int. J. LCA.

Online-First is not a pre-print service—the publications are in their final form; they can be neither changed nor withdrawn. However, an Erratum can be added. For the publication in [the printed] Int J Life Cycle Assess, only the final page numbers, the citation line, and the online publication date will be added.

Citation of Online-First articles: They can be cited by the Digital Object Identifier (DOI) which is an identification code and included in both the print and the electronic versions. The corresponding URL is listed in the abstracts of Online-First articles. Once a DOI is assigned to an article, it accompanies the paper until its final fate and should therefore be part of the citation line of this article. The DOI secures the identification of online articles wherever they are stored.

… The Online-First directory hosts the accepted articles before they are printed. Then they are shifted into the data bank [the directory of the online editions] where they can be identified by … the DOI" (Heinrich and Klöpffer 1999).

In November 1999, the first five OnlineFirst papers were published, only one year after the introduction of the DOI by ecomed.

Despite the growing interest in moving towards online publishing, much attention continued to be paid to the printed edition, above all to the cover pictures. From issue to issue it was a great pleasure to create the cover pictures together with Edwin Grondinger (abavo, Buchloe, Bavaria, Germany), an exceptionally skilled designer. The cover pictures always referred to one or several papers of the individual issues. Grondinger was briefed on the underlying idea by the Managing Editor, but additionally contributed his own vision and imagination, and the final product convinced authors and readers.

Figure 5.1 shows a beautiful example, namely the 'Schlossberg' at Graz, Austria (issue no. 4, vol. 2, 1999). This cover picture refers to a conference report on the

Fig. 5.1 Cover picture of vol. 2, no. 4, 1999

Society of Automotive Engineers (SAE), 'LCA on the Third SAE Conference on Total Life Cycle in Graz, Styria, Austria'.

Authors of specific articles were at times invited to suggest cover pictures. One of the most impressive results was the cover of the special issue no. 1, vol. 10, 2005, on the occasion of the five-year existence of the ecoinvent database (see Sect. 7). Rolf Frischknecht, the Special Issue Editor and the ecoinvent project manager, provided the fascinating idea for this cover picture (Fig. 5.2).

In 2008, the journal was transferred to Springer-Verlag (Fig. 5.3).

Fig. 5.2 Cover picture of vol. 10, no. 1, 2005

5 The National Societies

5.1 LCA Society of Japan

LCA was formally recognised in Japan with the creation of the industry sponsored Japan LCA Forum in 1991. In 1995, the LCA Society of Japan (JLCA) was established and funded by MITI (Ministry of International Trade and Industry). The society included over 400 members from material, energy, construction and distribution companies, as well as from the educational and public sectors. JLCA published a quarterly newsletter 'Forum News' in Japanese.

Fig. 5.3 Cover picture of vol. 6, no. 7, 2007, announcing the transfer to Springer

The Corner of the LCA Society of Japan (JLCA) in Int J Life Cycle Assess was established in vol. 2, no. 2, 1997; giving regular coverage of JLCA activities. In turn, JLCA previewed and profiled the journal in ECP Network News from Japan (ECP—Environmentally Conscious Products). ECP, the Environmental Network Newsletter, was sponsored by JEMAI, Japan (JEMAI—Japan Environmental Management Association for Industry).

The cooperation between Int J Life Cycle Assess and JLCA was a symbiotic relationship, or in modern speak, a win-win situation which proved to be fruitful. Other societies soon followed.

The JLCA was divided into three groups:

- examination of LCA methods
- examination of database construction
- application of LCA

The Society had two main objectives:

- the exchange of information and
- the establishment and use of common LCA data.

In 1998, the National LCA Project, namely the development of a database, started. It was planned by JLCA and financially supported by MITI. The five-year project consisted of members from industry, university and research institutes. The main subject areas were: Development of an Inventory Data Base; LCA Case Studies; LCA Application Guidelines; LCA Education and Propagation; Inventory Analysis Methodology; Impact Assessment Methodology (Morimoto 1997).

This National LCA Project was covered in several papers in Int J Life Cycle Assess:

- The Progress of the Impact Assessment Study Committee in the National LCA Project of Japan (Itsubo 1999a)
- The Progress of Inventory Study Committee WG2 in the National LCA Project in Japan (Itsubo 1999b)
- The Progress of the Database Study Committee in the National LCA Project of Japan (Nakahara 1999)
- Launch of the Damage Function Sub-Committee in the National LCA Project of Japan (Itsubo 2000)
- Current Activities of the National LCA Project in Japan (Yano et al. 2000)

In the following two years, research focused on various case studies and procedures of interpretation.

As of October 2007, JLCA had about 362 members including 43 industry associations, 3 other societies, 197 businesses, 68 individuals, and 51 university research organisations. About 1,000 registered users had accessed the JLCA database (Nakano et al. 2007).

A significant event in the cooperation with Japan was the publication of the special issue 'LCA in Japan' in 2000. Matthias Finkbeiner and Yasunari Matsuno were the editors.

Between 1993 and 1998, the Ecomaterials Project conducted systematic LCA studies. International conferences in 'Ecobalances' occurred biannually since 1994. "While the Japanese may not have been very active in the development of LCA *per se*, they have been at the forefront of the government-industry cooperation, specifically with the integration of life cycle concepts into decision making, reporting and public education" (Hunkeler et al. 1999).

"The conference has grown from the prior meetings in 1994 and 1996 which introduced life cycle assessment into Japanese academia and industry. The 159 papers presented from authors representing each major developed and developing regions of the globe, make the EcoBalance III Conference the largest LCA-related meeting in the world. The Third Conference focused on the application of life cycle assessment, as well as its associated fields of life cycle management, ecodesign and life cycle thinking, towards both practical industrial cases as well as national and supranational policy related issues. The sub-theme of the conference was movement towards sustainability and, as will be discussed, significant progress has been made to evolve the life cycle concept into a practical tool. Indeed, a primary conclusion of the conference was that a move towards industrial ecology would require a shift in the development of firm based assessment methods (Design for Environment, Ecomaterial Selection, Life Cycle Assessment) to those which could be oriented towards multi-stakeholders, specifically consumers, and related in terms of market parameters such as value. The rigorous methods, via the development of ecometrics, were seen as means for validating such tools" (Hunkeler et al. 1999).

With the review on the Third International Conference on Ecobalances (ICEB), Hunkeler et al. opened a complete documentation through the 9th ICEB in Int J Life Cycle Assess:

The Third International Conference on Ecobalances Movement towards sustainability
Tsukuba, Japan, November 25–27, 1998
David Hunkeler, Ryoichi Yamamoto, Itaru Yasui
Int J Life Cycle Assess (1999): 118–120

The Fourth International Conference on Ecobalances Methodologies for decision making in a sustainable twenty-first century
Tsukuba, Japan, October 31 to November 2, 2000
David Hunkeler
Int J Life Cycle Assess (2001): 49–51

The Fifth International Conference on Ecobalances Practical tools and thoughtful principles for sustainability
November 6–8, 2002, Tsukuba, Japan
Atsushi Inaba, David Hunkeler, Gerald Rebitzer, Matthias Finkbeiner, Claude Siegenthaler s and Konrad Saur
Int J Life Cycle Assess (2003): 1–5

The Sixth International Conference on Ecobalances Development and systematizing of ecobalance tools based on life-cycle-thinking
October 25–27, 2004, Tsukuba, Japan
Yasunari Matsuno1, Norihiro Itsubo, Shigeyuki Miyamoto, Toshiharu Ikaga, Hiroki Hondo and Atsushi Inaba
Int J Life Cycle Assess (2005): 159–162

The Seventh International Conference on EcoBalances Designing our future society using systems thinking
November 25–27, 2006, Tsukuba, Japan
Hiroki Hondo1, Koji Tokimatsu, Tsuyoshi Fujita, Yasunari Matsuno, Michiyasu Nakajima, Kenichi Nakajima and Yuichi Moriguchi
Int J Life Cycle Assess (2007): 66–69

The Eighth International Conference on EcoBalances The challenge of creating social and technological innovation through system-thinking
December 10–12, 2008, Tokyo, Japan
Kenichi Nakajima and Yasunari Matsuno
Int J Life Cycle Assess (2009):577–583

The Ninth International Conference on EcoBalances Towards and beyond 2020
November 9–12, 2010, Tokyo, Japan
Keisuke Nansai, Yuki Kudoh, Hiroki Hondo, Kiyotada Hayashi, Kazuyo Matsubae, Kenichi Nakajima, Shinsuke Murakami, Masaharu Motoshita, Seiji Hashimoto, Minako Hara, Michiyasu Nakajima, Rokuta Inaba, Yasunari Matsuno, Yoshikazu Shinohara
Int J Life Cycle Assess (2011): 478–487

The Tenth International Conference on EcoBalances Challenges and solutions for sustainable society
November 22–23, 2012, Yokohama, Japan
Toshiharu Ikaga, Keio University (Chair), Shigeyuki Miyamoto, NEC Corporation (Vice chair), Hiroki Hondo*, Yokohama National University

As in previous years, the 10th ICEB attracted many participants from all over the world working in academia, industry and government. The conference had 200 people from Japan and 103 people from 22 overseas regions and nations, including: Korea, Germany, Taiwan, USA, Thailand, Switzerland, China, Finland, Indonesia, Norway, Sweden, The Netherlands, France, India, Italy, Australia, Brazil, Denmark, Ghana, Malaysia, Philippines, and Vietnam.

A further step in the close cooperation between Int J Life Cycle Assess and Japan occurred in 2007 with the introduction of a new corner to report the research activities of the Institute of Life Cycle Assessment, Japan (ILCAJ) which was founded in October 2004 (Matsuno et al. 2013).

"The goal of ILCAJ is to promote academic activities related to life-cycle thinking and to share expert knowledge with colleagues from wide-ranging backgrounds. Professor Ryoichi Yamamoto, University of Tokyo, has taken responsibility as Chairman of ILCAJ.

In April 2005, ILCAJ has successfully established its publication organ (in Japanese), The Journal of Life Cycle Assessment, Japan (J LCA Jpn). The issues appear every three months. J LCA Jpn publishes peer-reviewed research articles, commentaries and discussions, (technical) reports, lecture notes, presentations of research groups in Japan, among other" (Matsuno and Kondo 2007a, b).

Through 2010, abstracts of research articles as well as of commentaries and discussions published in J LCA Jpn were simultaneously published in Int J Life Cycle Assess, Corner: J LCA Jpn (Matsuno and Kondo 2008a, b).

5.2 *Indian Society for LCA (ISLCA)*

The Indian Society for LCA (ISLCA) was founded in December 1997 by NEEF—National Ecology and Environment Foundation Trust, Mumbai, India (http://www.neef.in/islca.html) (Sharma 1999, 2000).

The objectives of ISLCA include the following:

- Capacity building for the development of LCA in India through its courses, training programmes, conferences, seminars, research projects, etc,
- Integrating socio-economic concepts in LCA,
- Representing India in national and international forums on LCA and related areas,
- Networking with leading professionals in LCA and related fields,
- Promoting publications of ISLCA including its planned periodicals and newsletters, publications, videotapes, discs and other communication media.

Since 2000, Int J Life Cycle Assess has been the official publication organ of ISLCA, This means that the journal publishes a regular column about plans and activities of ISLCA (contact: Vinod K. Sharma). However, the ISLCA Corner rarely appeared, and the documentation stopped with issue no. 1, vol. 9, 2004, p. 69 (State of Environmental Product Declarations in India). Nevertheless, officially the journal still has a publication agreement with ISLCA, and NEEF has linked the journal at http://www.neef.in/index.html (link to: http://www.springer.com/environment/journal/11367).

5.3 Korean Society for LCA (KSLCA)

The Korean government introduced ISO 14040 (Environmental Management—Life Cycle Assessment—Principles and Framework) and ISO 14041 (Environmental Management—Life Cycle Assessment—Goal and Scope Definition and Inventory Analysis) as national standards (KS) and founded the Korean Society of Life Cycle Assessment (KSLCA) in 1997.

KSLCA published 3–4 newsletters and one 'technical' journal annually. The first issue appeared in 1999. Furthermore, the 2nd of the planned annual 'academic' conferences of KSLCA was held in 1999, with about 200 participants from academia and industry (Tak Hur 1999, 2000).

KSLCA was divided into four categories:

1. Policy and strategy,
2. Methodology development,
3. Database construction,
4. Case studies.

The Ministry of Commerce, Industry and Energy supported KSLCA for the development of public databases, and a 5-year national research project for the construction of national LCI databases (1998–2003) was established.

In 2003, the activities under the formal responsibility of KSLCA seemed to be on a good way (Tak Hur 2003). However, shortly thereafter, the documentation from and about the society ceased as well as contact with Prof. Tak Hur, indicating that the society ceased activity.

Outside of KSLCA Tak Hur published in 2009, together with Ik Kim, an article on 'Integration of working environment into life cycle assessment framework' (Kim and Hur 2009).

5.4 Australian LCA Society (ALCAS)

In 2001, the Australian LCA Society (ALCAS) was established (http://www.alcas.asn.au/).

"The aim of ALCAS is to promote and foster the responsible development and application of LCA methodology in Australia and internationally, with a view to contribute to 'Ecological Sustainable Development (ESD)' and to represent the Australian LCA community in the international arena. This will be achieved through:

- developing a national competence in LCA,
- fostering links with the international LCA community,
- organising a regular LCA Roundtable to facilitate information exchange and discussion on LCA amongst stakeholder groups,
- contributing to national policies, positions and approaches on LCA and its applications,
- increasing education and awareness of LCA among stakeholders including industry, academia, government, nongovernment organisations, LCA practitioners, end users and the general public" (Grant et al. 2001).

Before ALCAS was officially established and the ALCAS Corner founded in Int J Life Cycle Assess, Tim Grant (RMIT, Melbourne Victoria, Australia) reported on the 2nd National LCA Conference 'Moving from Problems to Solutions' (Grant 2000). In the meantime, the 8th conference has been planned on LCA and Carbon Footprinting, 'Pathways to Greening Global Markets' (16–18th July 2013, Sydney NSW, http://conference.alcas.asn.au/).

The ALCAS Corner was active only through 2004 (Grant et al. 2001, Editorial; Grant 2002; Grant and James 2002; James and Grant 2002; James 2003; James et al. 2003; James and Narayanaswamy 2004), but outside the Corner Karli James promoted publications on LCA activities in Australia (Foley and Lant 2009; James et al. 2002; May and Brennan 2003; Parsons 2007, 2010; Peters et al. 2010; Puri et al. 2009; Verghese et al. 2010; Ximenes and Grant 2013).

In March 2013, Barbara Nebel, the regional editor of LCANZ in Int J Life Cycle Assess (see Sect. 5.5 below) agreed to also represent ALCAS.

5.5 Life Cycle Association of New Zealand (LCANZ)

Established in 2004 as an informal network group, LCANZ (Life Cycle Association of New Zealand) (http://www.lcanz.org.nz) joined the societies in Int J Life Cycle Assess in 2005. The first documentation was published in the same year as the First

LCA Workshop/Roundtable, Rotorua, NZ, in February 2005, with about a dozen LCA practitioners from research organisations, universities and consultancies presenting overviews on current projects (Nebel and Nielsen 2005).

The Second LCA Workshop/Roundtable, Rotorua, NZ, took place exactly one year later in February 2006 (Nebel 2006). The group discussed the need for a more formal LCA platform in New Zealand.

In 2009, LCANZ was officially established, with Barbara Nebel as president (Dr. Barbara Nebel, Wellington, New Zealand), to provide a focal point for Life Cycle Assessment and Management work conducted in New Zealand.

In 2010, the First (official) New Zealand Life Cycle Assessment Conference took place, jointly organised by LCANZ and the New Zealand Life Cycle Management Centre (NZLCM) Centre. The title was 'Bridging the Gap between Tools and Practice'.

The theme for the Second New Zealand Life Cycle Assessment Conference in 2012 was 'Life Cycle Assessment: A Business Compass for Sustainable Development'. This theme reflected the increasingly important role that LCA plays in guiding and shaping business operations, management practices and strategies in New Zealand.

The main objectives of LCANZ include the following:

- Provide coordinated input to the New Zealand government on its policy development for matters relating to LCA/LCM, with a view to ensuring that government is advised of current work and the views of LCA practitioners
- Identify, prioritise and address barriers to widespread uptake of LCA/LCM, including: gaps in NZ LCA/LCM expertise, gaps in data
- Provide input into relevant standards and guidelines (national and international), where it is deemed appropriate to do so
- Review whether there is a need for professional recognition of LCA/LCM practitioners
- Facilitate access to relevant LCA/LCM experts in New Zealand
- Periodically review the need for LCA/LCM resources, and where a need is identified, facilitate the development of its/their production
- Provide technical advice on LCA/LCM work, where appropriate and feasible

5.6 *Other LCA Organisations and Networks*

5.6.1 SPOLD—Society for the Promotion of Life Cycle Development

SPOLD (Society for the Promotion of Life Cycle Development) started in 1992 originally as an industry association. However, other associate members (LCA consulting firms, universities, scientific institutes, etc.) later joined SPOLD.

SPOLD was a Brussels-based society created to promote the development and application of LCA. The organisation funded LCA-related scientific research, methodology development, education and communication. SPOLD brought together scientists from industry, consultancy and academia.

SPOLD identified two priority development areas to further improve the use of LCA:

1. To facilitate the public availability of life cycle inventory data, with a consistent format and with well documented data quality characteristics. SPOLD did not intend to create a new LCI database. Instead, the SPOLD database project collected all data available from different LCA stakeholders, within a consistent methodological framework. SPOLD recognised that the first task was to ensure the development of a consistent format for all the data that are available.
2. To create a constructive dialogue between industry, government, academic institutions and environmental groups to ensure broad alignment on the specific role of LCA for use in environmental decision and policy making.

SPOLD developed a common format for reporting LCI data to improve the transparency and comparability of LCI data and represent an important first step towards the establishment of a common LCI database (Singhofen et al. 1996).

As well as developing the format, SPOLD published a directory of sources of LCI data, as a first step in facilitating access to the data that were available (1995). This directory included information not only on the many reports and commercial software packages which contain LCI data, but also on the numerous data gathering initiatives which were completed or were underway under the sponsorship of industry, trade associations and national authorities (Bretz 1998; Hindle and de Oude 1996).

In 1995, SPOLD launched the project 'Winning Acceptance', to create a constructive dialogue and build consensus between the stakeholder industries, governments, environmentalists, professional groups, and academic institutions (see also SustainAbility Ltd.'s LCA Sourcebook (1993) that presents basic information on LCA techniques, practitioners, and data sources).

1998 was the last year of SPOLD's traditional role as industry's LCA sponsoring organisation. The reasons for its termination are manifold.

1. "The success of the SPOLD data format has out-grown the narrow limits of a traditional industry association" (Bretz 1998).
   The open source ecoSpold data format v1 (ecoSpold (v1)) was launched in 2000. Bo P. Weidema (http://www.lca-net.com) was instrumental in the development of the SPOLD LCI data format and database network from 1995 to 2001. Later he was executive manager of the ecoinvent database (2008 to 2012).
   The ecoinvent Centre was the first to use this data format for their own LCI database. Other databases adopted the format, and all important LCA software tools had an interface to use datasets in ecoSpold format.
2. In 1998 SETAC-Europe formed a work group on 'data availability and data quality' which comprised practically all members of SPOLD's work group 'Promoting Sound Practices', together with representatives from most important LCA data and software suppliers, and many other interested parties (Hischier et al. 2001).

3. Another issue of SPOLD, the integration of LCA into a comprehensive 'environmental toolbox' (together with MFA—Material Flow Analysis, ERA—Environmental Risk Assessment, etc.), was adopted by other organisations (LCANET and later CHAINET). The challenges of using LCA in small and medium enterprises are closely linked to those of gaining eco-efficiency improvements. The task of finding ways to stimulate and help companies in these important areas was beyond the resources of SPOLD.

5.6.2 LCANET—European Network for Strategic Life-Cycle Assessment Research and Development. A Strategic Research Programme for Life Cycle Assessment

DGXII at the European Commission subsidised a concerted action in the Environment and Climate programme for the establishment of a 'European Network for Strategic Life-Cycle Assessment Research and Development': LCANET. The task of this network was to describe the state-of-the-art of LCA methodology and to provide input to the EU Environment and Climate research and development programme.

The final document for the concerted action LCANET provided a programme for LCA research priorities in order to ensure more widespread use of LCA. It included meetings, workshops and intermediate reports. The result was the identification of the four following research themes:

1. Toolbox for life cycle assessment, including simplification, robustness, expert systems for filling data gaps and relationship to other tools
2. Decision making processes
3. LCA method development
   1. System modelling
   2. Characterisation
   3. Weighting
4. Uncertainty in all phases of LCA and the validation of software

LCANET reflected research needs as guidance to the EU 5th Environment and Climate framework programme. Int J Life Cycle Assess published a short version of the final document (Wrisberg et al. 1997) as well as a short version of the overall preface that substituted the tentative prefaces of the four reports (Udo de Haes and Wrisberg 1997a).

By 1997, LCANET had nearly 200 members. The results were supported by a wide community of European experts. Publications of LCANET results have given input to the research programme of the second phase of the 4th framework programme and to the fifth framework programme of the EU-DGXII Environment and Climate programme.

"However, it can be expected that the results will be of wider significance and will also, in a more general way, stimulate and focus LCA research and development in the forthcoming years. It is also to hope that the work of the network can be continued" (Udo de Haes and Wrisberg 1997, *Chairman and Coordinator of LCANET*).

The complete version of the 'Final Document for the Concerted Action LCA-NET' appeared as Volume 1 of the book series 'LCA Documents' (Udo de Haes and Wrisberg 1997b).

5.6.3 CHAINET—European Network on Chain Analysis for Environmental Decision Support

CHAINET was an EU-supported Concerted Action in the Environment and Climate Programme. Similar to its predecessor I.CANET, it was a European network and addressed the use of a variety of environmental tools. The concerted action commenced in December 1997 and had a duration of two years. Helias A. Udo de Haes and Nicoline Wrisberg continued their roles as Chairman and Coordinator (Klöpffer 2004; Udo de Haes and Wrisberg 2002; Wrisberg 1998).

The Aims

- linking the different scientific tool communities, problem owners and stakeholders,
- establishing a toolbox for chain analysis,
- investigating how tools can be applied in three selected cases to suggest specific directions for design and development.

The Cases Three cases were selected as vehicles to be useful for discussions on how tools can be applied in order to get information on environmental improvements. These are the supply chain, the use chain and the waste management chain for

- automobiles,
- electronic consumer goods,
- domestic washing of clothes.

The working groups, one for each case, included:

- identify environmental problems in the chain,
- describe the results from existing environmental analyses,
- discuss relevant tools for the analysis of environmental impacts,
- formulate guidelines for the application of tools.

The final products of the project were:

The guidebook provides a toolbox for chain analysis, linking demand for environmental information with supply of relevant information. In addition it gives information on the application of the toolbox in the three cases indicating specific directions in design and development.

The network consists of environmental problem owners (stakeholders) and tool experts.

5.6.4 ISOLP—International Society for LCA Practitioners

Discussions within the board of SPOLD and at the LCANET workshop in Nordwijkershout, The Netherlands, identified a need for an organisation for LCA practitioners who did not seem to be fully represented and supported by the science-based organisations of SETAC, SETAC Europe, SPOLD, or ISO. "The [ISO] conventions which are required for allocation and similar topics … cannot be solved through science alone, but instead demand a social affirmation. A society representing the majority of practitioners would have the mandatory authority for proposing conventions which could finally enter into future improvements of the ISO 14040 series. Neither science nor industry and standardization bodies alone are able to do this. Aside from these and similar technical questions, there are other important questions of ethics and sponsor-practitioner relationships which affect all LCA practitioners throughout the world and should therefore be addressed by this prospective society" (Klöpffer 1997, Editorial).

This Editorial was supported by Laurent Grisel, Ecobilan France and Bo Weidema, IPU Denmark. It invited open discussion but did not get immediate feedback.

There were some indirect responses in 2001 and 2002 (Klöpffer and Heinrich 2001, 2002; Heinrich and Klöpffer 2002), but the actual response occurred 15 years later via the Editorial of Martin Baitz et al. (2013). It is not a direct reply to founding a society for LCA practitioners, but it picks up the discussion on how to improve cooperation in the use of LCA in both theory and practice. "The authors share the implications of LCA in daily businesses and practice and aim to nurture and strengthen the interfaces between scientific findings and application. Working together to encourage a broader application of 'good practice' LCA in industry as well as strengthening scientific LCA work towards 'applicable science' will develop and reinforce professional LCA work and technical implementation in the academic and business arena. This article is written with a primary focus on industrial applications and research in applied science and with less emphasis on specific governmental applications" (Baitz et al. 2013).

Walter Klöpffer commented on this editorial and suggested the following solution: "After publication of a new method … a further step should follow: a broad testing with real product systems. This should be done by the practitioners and financed either by industry associations or governmental and supragovernmental organizations" (Klöpffer 2013).

5.6.5 UNEP/SETAC Life Cycle Initiative

The Initiative (http://lcinitiative.unep.fr/) responds to the call by Governments around the world for a Life Cycle economy in the Malmo Declaration (2000). It contributes to the 10-Year Framework of Programmes to promote sustainable consumption and production patterns, as requested at the World Summit on Sustainable Development in Johannesburg (2002).

Due to the complementarity of the journal and the Initiative, the board of the UNEP/SETAC Life Cycle Initiative decided in 2003 to establish an official collaboration with Int J Life Cycle Assess to become the **Associated Journal of the UNEP/SETAC Life Cycle Initiative**.

As part of the collaboration, the journal agreed to regularly inform readers about recent developments and activities of the Life Cycle Initiative and to provide active members of the Initiative from developing countries the journal for a reduced fee.

Already before this agreement, the journal published special issue on the launch of the UNEP/SETAC Life Cycle Initiative (no. 4, vol. 7, 2002). The launch took place on 28 April 2002 during UNEP's 7th High-level Seminar on Cleaner Production, and in presence of the former SETAC President Lorraine Maltby and UNEP's Executive Director Klaus Toepfer. The latter prepared an editorial for the journal and thanked its editor-in-chief, Walter Klöpffer, not only for his valuable work in promoting Life Cycle Assessment and Life Cycle Management on an international level, but also for his support of the Life Cycle Initiative by this special issue (Toepfer 2002).

Since 2003, Int J Life Cycle Assess has been continually reporting on the Initiative's activities in the 'Corner of the UNEP/SETAC Life Cycle Initiative': for example in 2005 about progress in Life Cycle Impact Assessment within the UNEP/SETAC Life Cycle Initiative (Jolliet et al. 2005), in 2007 about the first phase 2 activities of the Initiative (Sonnemann and Valdivia 2007) and in 2011 about the process on global guidance for LCA databases (Sonnemann et al. 2011).

Furthermore, the journal has published relevant deliverables such as the LCIA Midpoint-damage Framework of the UNEP/SETAC Life Cycle Initiative in 2004 (Jolliet et al. 2004), the activity of Task Force 1 on global life cycle inventory data resource (Curran 2006) and a special issue on USEtox in 2011 (Hauschild et al. 2011).

The journal has also been helpful in announcing conferences such as CILCA (International Conference on Life Cycle Assessment) in Costa Rica in 2005 (Sonnemann 2005), the recent Indian life cycle assessment and management conference in 2012 (Datta et al. 2012; Datta and Valdivia 2013) and the United Nations Conference on Sustainable Development (Rio+20) in 2012 (Valdivia et al. 2012) as well as in reporting on events such as in the form of key observations arising from papers on sustainable production, use and recycling of natural resources from the symposium in Portland in 2006 (Fava et al. 2006).

The sponsors of the Initiative expect this fruitful cooperation to continue in the future. As a first step, updates on recent developments in Life Cycle Impact Assessment and the finalisation and current dissemination activities of the publication on global guidance principles on LCA databases are foreseen. Moreover, special issues on Life Cycle Sustainability Assessment and global land use impacts on biodiversity and ecosystem services in LCA are under preparation.

Guido Sonnemann and Sonia Valdivia report on the Initiative in Chapter 4, this volume.

5.6.6 Swiss Discussion Forum on Life Cycle Assessment

In 2005, the Swiss Discussion Forum on Life Cycle Assessment started to report on the individual sessions in Int J Life Cycle Assess.

Mission and Organisation The Discussion Forum on Life Cycle Assessment (http://www.lcaforum.ch/) applies to practitioners from industry, consulting companies and administration and to LCA scientists, from Switzerland and abroad. Each LCA forum is dedicated to a specific topic of immediate interest related to

- experiences and challenges with LCA application in industry and administration,
- scientific questions in life cycle inventory and life cycle impact assessment methodology development,
- dissemination of new scientific findings and results of relevant LCA studies.

Each forum offers an 'open floor' session for short presentations. The topics are defined by the advisory board. Proposals are welcome.

Advisory Board

- Dr. Yves Loerincik (president)
- Dr. Arthur Braunschweig
- Norbert Egli
- Dr. Rolf Frischknecht
- Dr. Gérard Gaillard
- Prof. Stefanie Hellweg
- Roland Hischier

Peer Review/Critical Review—23rd LCA Discussion Forum The range of topics is broad (Braunschweig 2005; Doublet and Jungbluth 2011; Friot et al. 2005; Frischknecht and Flury 2011; Frischknecht et al. 2009; Loerincik et al. 2005; Saner et al. 2012; Schuerch et al. 2012; Siegenthaler and Margni 2005).

Coverage in Int J Life Cycle Assess began with the topic 'Quality Control and Peer Review' by Arthur Braunschweig (2005). Braunschweig referred to the paper 'The Critical Review Process According to ISO 14040: An Analysis of the Standard and Experiences Gained in its Application' by Walter Klöpffer (2005, the same issue and presented at the 23rd LCA Discussion Forum which took place at ETH Zurich on 23 September 2004).

'Peer Review' or 'Critical Review' has been a perennial problem from 1993 through today. A peer review for LCA-studies was first proposed in the SETAC guidelines 'A Code of Practice' (1993). SETAC recommends "the accompanying or interactive critical review, which should be preferred, over the review '*a posteriori*', which offers considerable risks in regards to the duration and costs of an LCA study" (Klöpffer 2005). In contrast, ISO 14040 describes the three methods but does not recommend any of them, which Braunschweig criticises "The forum expressed its hope that the current revision of 14040 will not increase ambiguities, but rather clarify such issues in a reasonable way" (Braunschweig 2005). From personal experience, Klöpffer leans toward supporting the recommendation by SETAC (Klöpffer 2005).

The current revision of 14040 (1997) addressed by Braunschweig (above) concerns ISO 14040: 2006 and 14044:2006 (Environmental Management—Life Cycle Assessment—Requirements and Guidelines). ISO 14040 is a framework and guidance standard, while ISO 14044 contains all technical requirements and guidelines thereon. ISO 14040:2006 and ISO 14044:2006 are the core standards of LCA. However, the contradiction between the title of Sect. 7.3.3 in ISO 14040:1997 (Critical review by panel of interested parties) and the content (this panel *may* also include other interested parties) has not been removed in the new standard of 2006. The inclusion of interested parties is again described as optional ('may' and not 'shall'). "I recommend to all commissioners of comparative LCAs to install interactive rather than 'a posteriori' critical review" (Klöpffer 2012).

In the meantime (2013), ISO TS 14071 is in development (Life cycle assessment—Critical review processes and reviewer competencies—Additional requirements and guidelines to ISO 14044:2006) and may propose how to proceed in practice.

5.6.7 LCA Activities in Spain, Italy and Greece

Spain APRODACV (Asociación Española para la Promoción del Desarrolo des Análisis del Ciclo de Vida), the Spanish Association for the promotion of LCA development, was established in 1995. It appealed to academia, industry, consulting, administration (Domenéch and Fullana 1996). The first LCA workshop in Spain, 'LCA 2000', took place (Verger 1997), and the first LCA book in Spanish appeared by Pere Fullana and Rita Puig and was reviewed by Michael Hauschild (1998).

Italy Around 1998, only the major industrial companies used LCA methodology in Italy. The main limitation to the expansion of LCA activities in small and medium companies resulted from the important investment needed (Giacomucci and Baldo 1998). However, in 1999 Giacomucci and Baldo reported on first experiences with LCA certification according to ISO 14040 (1999). Simultaneously, the LCA Society of Italy (Associazione Italiana di Analisi del Ciclo di Vita) was established (Baldo and Giacomucci 1999).

In 1997, the Italian Environmental Protection Agency ANPA (Agenzia Nazionale per la Protezione dell'Ambiente) promoted the construction of a database, called I-LCA, to support Italian LCA practitioners. The first version of I-LCA was available in 1999, and at the beginning of 2000, the second version appeared which was supported by three consulting companies: Ambiente Italia, Boustead Consulting Ltd., UK and Ecobilan, France (Baldo and Pretato 2001).

As the LCA Society of Italy failed, ENEA (Italian National Agency for New Technologies, Energy and the Environment), supported by the Ministry of Environment, promoted and coordinated the informal Italian Network on LCA—Associazione Rete Italiana LCA (Cappellaro et al. 2008, editorial).

This network runs a technical secretariat managed by ENEA, a website (www.reteitalianalca.it) and a newsletter. It organised several workshops and conferences, and finally, on June 2012, a not-for-profit scientific association was established with Paolo Masoni as president.

Greece On December 16th, 1997, the first Greek workshop on LCA was organised at the Aristotles University of Thessaloniki by the Laboratory of Heat Transfer and Environmental Engineering (LHTEE) (Moussiopoulos and Koroneos 1998a).

The Hellenic Life Cycle Assessment Network (HELCANET) was created in February 1998 by LHTEE of the Aristotle University of Thessaloniki (AUT) (Moussiopoulos and Koroneos 1998b). LHTEE is one of the first Greek bodies to get involved in LCA activities.

An impressive overview on the LCA Activities in Greece was given by Boura et al. (2000). This was the last contribution from Greece on LCA activities, but a number of scientific papers followed.

The website presents the objectives of HELCANET (http://aix.meng.auth.gr/lhtee/index.html):

- to promote and support scientific research, education, training, dissemination of information and development in the area of life cycle issues,
- to catalyse the development and application of life cycle assessment by pooling the talent and resources of industry and other organisations interested in LCA,
- to be a platform for discussion on LCA research and development by regular and rapid exchange of information between Greek universities, research institutes, companies, authorities and governmental organisations.

HELCANET focuses on social dialogue and LCA methodology development in Greece, the piloting of product and process LCA (waste management, energy systems, building materials), on ecolabelling criteria, ISO 14040, inventory, databases, data quality, impact assessment, recycling, policy, design for environment.

6 Topics and Subject Areas

As in any developing research field, new topics appear and supplement the already established ones. In 2009, the following topics and subject areas were identified and attributed to subject editors, many of who are still active in the same position (Klöpffer and Heinrich 2009):

- Carbon footprinting (Subject editor: Matthias Finkbeiner)
- Data availability, data quality in LCA (Subject editor: Martin Baitz)
- EU life cycle policy and support (Subject editor: David Pennington)
- Input–output and hybrid LCA (Sangwon Suh, Shinichiro Nakamura)
- Land use in LCA (Subject editor: Llorenç Milà I Canals)

- LCA for agriculture (Subject editors: Gérard Gaillard, Seungdo Kim)
- LCA for energy systems and food products (Subject editor: Niels Jungbluth)
- LCA of waste management systems (Subject editor: Shabbir H. Gheewala)
- LCIA of impacts on human health and ecosystems (Subject editors: Michael Z Hauschild, Rana Pant, Ralph K. Rosenbaum)
- Life cycle management (Subject editors: Gerald Rebitzer, Yasunari Matsuno, Wulf-Peter Schmidt, Thomas Swarr)
- Nontoxic impact categories associated with emissions to air, water, soil (Subject editor: Mark Huijbregts)
- Societal life cycle assessment (Subject editor: David Hunkeler)
- Uncertainties in LCA (Subject editor: Andreas Ciroth)
- Water use in LCA (Subject editor: Annette Koehler)
- Wood and other renewable resources (Subject editors: Joerg Schweinle, Barbara Nebel, Liselotte Schebek, Frank Werner)

In the following years, these issues were supplemented by Life Cycle Costing (LCC), and Life Cycle Sustainability Assessment (LCSA), among other topics.

6.1 Life Cycle Management

Life Cycle Management essentially embraces many applications of Life Cycle Thinking, product- as well as company-related LCAs and simplified methods not or not fully compliant with the ISO standards. The methods used may also go beyond (environmental) LCA and contain Life Cycle Costing (LCC) and Social Life Cycle Assessment (SLCA) as a basis for Life Cycle Sustainability Assessment (LCSA). This consideration of the 'three pillars' of sustainability is often called the 'triple bottom line' in industrial management.

The magic word in relation to LCM has been 'toolbox'. LCM uses a toolbox (i.e. several methods besides LCA such as LCC—Life Cycle Costing, DfE—Design for Environment) rather than just one well-defined method as is LCA. This allows a growing number of applications to the 'tool' LCA. This makes LCM attractive for small and medium sized enterprises and explains its success as a complement to LCA.

6.1.1 Editorial: 'How to Communicate LCA Results' by Walter Klöpffer and Almut B. Heinrich, Int J Life Cycle Assess 5(3): 125 (2000)

"How to communicate LCA results?" question Klöpffer and Heinrich the readership of the journal due to the following discussions:

1. Should methodological papers or case studies preferably be published?
2. Should the journal be divided into a fast part for news, discussions and practical applications in form of an electronic, supportive information section, and a part for scientific, peer-reviewed methodological articles?

Simultaneously, the series ISO 14040–43 was completed in March 2000. Authors and commissioners of LCA-studies had to be aware that 'comparative assertions' are only acceptable according to the international standard.

The result on this editorial was animated and controversial.

"While the editorial mostly discussed how to present LCA results it paid little attention to the question of *if* they should be published and, if so, for whom. First of all we have to recognize that it is not *a priori* clear that LCA results should be published in journals at all" (Hofstetter 2000).

"The first issue of the year 2000 contains two articles on methodology. Everything else involves valuable information for LCA-users and researchers. The Int. J. LCA must earn a good reputation and rating when articles of superior quality on methodology are seen to appear. Otherwise, the researchers will feel compelled to turn to other journals.

Accordingly, the Int. J. LCA could also appear in two parts: A 'more rapid' part for News and Discussions, and a 'slower' part for the scientific and reviewed articles" (Frischknecht 2000).

"The Int. J. LCA should endeavour to advance LCA in all its aspects. Let me support the editorial that the journal should not focus exclusively on LCA methodology. In my opinion, the essential barrier in using LCA within industry is not methodology, but the barrier is the continuing need for inventory data. To practice LCA means having the necessary data to cover all parts of the system, knowing the data's utility and uncertainty are adequate for the study's goal and scope, and making sure that the data are adequate and will fit the impact assessment methodology that you have chosen" (Owens 2000).

An event occurred that solved this problem. It was the **First International Conference on Life Cycle Management** in Copenhagen, August 27–29, 2001. This conference was initiated by Allan Astrup Jensen during the SETAC World Congress in Brighton, May 2000. Thereupon the integration of the new section 'Life Cycle Management' into the journal (2002) took place.

The LCM Conference 2001 was a fascinating event. It attracted approximately 270 participants. Plenary lectures were held on the first and third days of the conference with three parallel sessions on the second day. Fifty-three platform presentations were complimented by forty-seven posters. A special characteristic of LCM 2001 was that it attracted much more interest from businesses (multinational corporations as well as SMEs) than most conferences in this area. By the way, the CHAINET Toolbox and Network was the focus of two sessions (see Sect. 5.6.3 of this chapter). Historically, the aim of CHAINET was to broaden the scope of the preceding LCANET program, which focused on LCA. More specifically, the aim was to link demand and supply of environmental information in the field of LCM (Hunkeler et al. 2001).

There had been much discussion in the journal before the section 'Life Cycle Management' could be integrated.

6.1.2 Editorial: 'Two Planets and One Journal' by Walter Klöpffer and Almut B. Heinrich, Int J Life Cycle Assess 6(1) 1–3 (2001)

In their editorial 'Two Planets and One Journal', Klöpffer and Heinrich (2001) summarised the various expectations and demands which editors, authors and readers expressed toward the journal. The conceptions and ideas depended mainly on the planet they inhabited: the planet of the Method developers (Sect. A) and the planet of the Practitioners (Sect. B).

a. **Method developers** (Academic background)
   - High level methodological papers
   - Case studies should be published only if new methods are applied, to test the methods in real life ('Feed-back', not provided by grey literature)
   - Other case studies should be published in sector-specific journals or on the web
   - The results of such case studies are of no interest to people working in LCA, their target public is unclear
   - Separation of the Journal into a (rapid) newsletter and a slower high-level Journal of highest possible reputation and rating (Science citation index, etc.)

b. **Practitioners** (Industry)
   - The Journal should advance all aspects of LCA, not only methodology
   - The Journal should help to provide sources of inventory data, improved sources of public data!
   - The Journal should help provide the required data exchange format!
   - No objection against a well-balanced share of methodology papers, but only few people are interested
   - Case studies on a high scientific level should be the main focus, Life Cycle Management (LCM) gains importance
   - Method development comes to an end (ISO finished), but: new methods to progress science (agreement seems to be that Life Cycle Impact Assessment belongs to this group) should be published

On this basis, Klöpffer and Heinrich disseminated a questionnaire to all 53 editors of the journal in December 2000. 45 editors responded to this questionnaire with the following result:

1. Separation of peer-reviewed and non peer-reviewed contributions:
   76 % yes, 18 % no
2. Concentration on methodology papers:
   20 % yes, 69 % no
3. Concentration on case studies:
   4 % yes, 84 % no
4. Mixture of contributions (status quo):
   4 % yes, 22 % no
5. Separation of contributions within each issue:
   56 % yes, 27 % no
6. Full studies in the internet:

77 % yes, 11 % no

7. Life Cycle Management (LCM) as title or subtitle:
   38 % yes, 53 % no

"We interpret these results as approval for a multitude of contributions, especially for methodology papers and case studies. For the sake of scientific reputation, however, a clear separation of peer-reviewed and non peer-reviewed contributions seems to be desirable. The comments show that a separation in two journals, as was suggested previously …, is not considered to be a good solution. Electronic publishing of comprehensive LCA-studies is considered to be promising by three quarter of the respondents. The inclusion of LCM in the title or as a subtitle polarized the board more than the numerical result can show (53 % no, 38 % yes): the comments expressed both clear refusal and enthusiasm. We shall observe the development and decide later about using LCM as a sub-title. Meanwhile, the electronic 'Gate to Environment and Health Science (Gate to EHS)', Section 'Life Cycle Management', stands wide open for peer-reviewed contributions on LCM" (Klöpffer and Heinrich 2001).

6.1.3 LCM in the Internet-Journal 'Gate to Environmental and Health Science (EHS)'[2] and the Discussion Forum 'Global LCA Village'

There was much motion and drive in the area of Life Cycle Management (Klöpffer and Heinrich 2002).

- LCM has been established in several industrial companies striving for sustainable development via life cycle-based methods
- Several articles submitted to Int J Life Cycle Assess dealt with LCM
- A working group of SETAC Europe was established in 1998 to explore the use of LCA and similar instruments in actual industrial management practice
- An international LCM conference was initiated by Allan Astrup Jensen during the SETAC World Congress in Brighton, May 2000 and successfully took place in Copenhagen, August 2001 (Hunkeler et al. 2001). Another workshop which immediately followed the congress was initiated by UNEP and SETAC in order to enlarge the profile of the UNEP/SETAC Life Cycle Initiative (Sonnemann et al. 2001)

In responding to this development including the results of the questionnaire sent in 2000 (Sect. 6.1.2 of this chapter), Walter Klöpffer and ecomed publishers[3] offered the online discussion forum 'Global LCA Village' and the area 'Life Cycle

[2] The aim of the stand-alone Internet Journal 'Gate to Environmental and Health Science (EHS)' was to expand the Scientific Journals at ecomed publishers, before they were transferred to Springer-Verlag (The International Journal of Life Cycle Assessment, Environmental Science and Pollution Research, Journal of Soils and Sediments, Umweltwissenschaften und Schadstoff-Forschung).

[3] ecomed was the publisher of the 'Scientific Journals' before they were transferred to Springer-Verlag (2008).

Management' in 'Gate to EHS' (Environmental and Health Science). 'Global LCA Village' was well-accepted and addressed LCA researchers, practitioners and industrialists. The access was free. 'Gate to EHS' was a stand-alone Internet Journal with restricted access.

The following sections in the LCM area were developed in 'Gate to EHS': Management Systems and Auditing (ed.: Matthias Finkbeiner), Life Cycle Costing (eds: David Hunkeler and Gerald Rebitzer), and Design for Environment (eds: Wulf-Peter Schmidt and Thomas E. Swarr).

Both publications ceased with the transfer of the 'Scientific Journals' (see footnote 5) to Springer-Verlag (2008). The contributions cannot be accessed anymore, not even by the DOI of the individual articles.

6.1.4 Editorial: 'LCM—Integrating a New Section' by Almut B Heinrich and Walter Klöpffer, Int J Life Cycle Assess 7(6): 315–316 (2002)

It was a difficult decision to integrate an LCM section into the journal. "In May 2002, we were still of the opinion that Int J LCA was not the optimal place for papers on LCM …. In the meantime [half a year later] we think that the journal (the printed Int J LCA) should acknowledge the developments in the LCM area more strongly. LCA, however, will remain the clear focus. LCM reflects the remaining needs that LCA alone cannot satisfy; therefore, it may be regarded as a second-generation development" (Heinrich and Klöpffer 2002).

In November 2002, Thomas E. Swarr defined LCM as such: "To me, LCM is the organising framework so we actually use the science of LCA to achieve improved performance. I find myself trying to communicate between specialists who are only comfortable with complex databases, and business executives who are only comfortable with PowerPoint bullet slides. We need a better balance between theory and practice" (Heinrich and Klöpffer 2002).

In 2006, Thomas E. Swarr referred to an LCM definition by Jensen and Remmen 2005[4]: "Life cycle management has been defined as the application of life cycle thinking to modern business practice, with the aim to manage the total life cycle of an organization's products and services toward more sustainable consumption and production" (Swarr 2006).

A specific LCM Editorial Board was responsible for the LCM papers:

- Matthias Finkbeiner, Germany
- David Hunkeler, Switzerland
- Yasunari Matsuno, Japan
- Gerald Rebitzer, Switzerland
- Wulf-Peter Schmidt, Germany
- Thomas E. Swarr, USA

[4] Jensen AA, Remmen A (2005): Background report for a UNEP guide to life cycle management, revised March, http://www.uneptie.org/pc/sustain/lcinitiative/publications.htm.

Simultaneously, four of them (Finkbeiner, Hunkeler, Matsuno, Schmidt) were regular members of the editorial board. Over the years, this separation between the regular and the specific editorial board was removed.

6.1.5 The LCM Conferences

Allan Astrup Jensen (the initiator of LCM 2001, Copenhagen, Denmark, see above), David Hunkeler, Gérard Gaillard, Stefanie Hellweg and Kim Christiansen reported on the second LCM Conference in Barcelona September 5–7, 2005 (Jensen et al. 2005). It featured some three hundred participants and was split into four parallel sessions with 125 oral presentations:

- Production systems,
- Agriculture & energy,
- Services,
- Integration tools.

During the conference over two hundred posters were exposed.

The overall message from the plenary lectures at LCM 2005 was a plea from industrialists to render LCA more relevant and applicable within a corporate context.

The third LCM Conference was held in Zürich, Switzerland, 27–29 August 2007, organised by Stefanie Hellweg and Gerald Rebitzer (see Sect. 7 of this chapter). UNEP/SETAC Life Cycle Initiative was associated. The conference discussed the theme 'From Analysis to Implementation'.

The fourth LCM Conference, LCM 2009, took place from 6–9 September 2009, in Cape Town, South Africa. The overall theme was 'The Global Challenge of Managing Life Cycles'. The conference was hosted by the University of Cape Town and supported by the United Nations Environment Program. The 180 delegates included 40 South Africans, 20 from other African countries, and 140 from as far afield as Brazil, Sweden, Japan, and Australia. This made LCM 2009 a truly global international conference.

LCM 2009 was successfully engaged with the critical questions of what it means to manage (not merely shift) the environmental and social impacts of global economic activity, what this entails for industry and public services in emerging economies, and how supply chains, networks, and partnerships can be stimulated and managed to deliver truly sustainable practice.

While the focus of the conference was LCM, LCA remains a main analytical tool for supporting LCM. This is clearly shown by the overall program in which roughly half of the contributions focused on or used LCA (Potting et al. 2010).

The fifth LCM Conference, LCM 2011, was held on August 28–31, 2011, in Berlin, Germany. Matthias Finkbeiner, Germany, was the chair and Stephan Krinke, Germany, the co-chair. The conference motto 'Towards Life Cycle Sustainability Management' addressed the challenge of implementing sustainability concepts. The conference featured 500 delegates, 180 presentations and 3 poster sessions (roughly 150 posters). The conference was documented by a separate volume 'Towards Life

Cycle Sustainability Management' edited by Matthias Finkbeiner and published by Springer[5].

The sixth LCM Conference, LCM 2013, was held in Gothenburg, Sweden, August 25–28, 2013.

The seventh LCM Conference, LCM 2015, will take place in Bordeaux, France.

First Indian Life Cycle Assessment and Management Conference 2012 A new conference series in India was initiated by the Federation of Indian Chambers of Commerce and Industry and UNEP/SETAC Life Cycle Initiative. They organised the First Indian Life Cycle Assessment and Management Conference (ILCM 2012) on 21–23 August 2012 in New Delhi, India (Datta et al. 2012).

The aim of ILCM is the application of tools for guiding governments, consumers and business towards a sustainable quality of life in India.

The key area of the conference concerned life cycle approaches regarding

- Methodology, standards, databases, etc.,
- Sustainable production,
- Sustainable consumption,
- Policy goals.

The organising team was:

- Archana Datta (archana.datta@ficci.com)
- Philip Strothmann (philip.strothmann@unep.org)
- Sonia Valdivia (sonia.valdivia@unep.org)
- Bruce Vigon (bruce.vigon@setac.org)

Second Indian life cycle assessment and management conference 2013: Creating business value through sustainable strategies (Bangalore, India, 26–27 September 2013)

Deriving inspiration from Rio+20 Sustainable Development dialogues and suggestions of the ILCM 2012 delegates, ILCM 2013 aims to showcase practical examples. Therefore, ILCM 2013 invites policy makers, business managers and social scientists.

Topics:

- Life Cycle Approaches: Local vs global perspectives
- Life Cycle Approaches: Business opportunities and challenges in using LCA
- Life Cycle Approaches: Social LCA for developing the institutional framework in India

[5] The table of contents can be downloaded at http://download.springer.com/static/pdf/978/bfm%253A978-94-007-1899-9%252F1.pdf?auth66=1362846876_2e42d6292899f2eea084362378e7b1e4&ext=.pdf.

6.2 *Life Cycle Costing (LCC)*

According to Hunkeler and Rebitzer (2003), Life Cycle Management is a business toolbox involved in product- and firm-based decision-making. (Environmental) Life Cycle Costing (LCC) is part of this toolbox.

"With LCC being a major component of the new LCM section in the Int J LCA (Heinrich and Klöpffer 2002) we, the editors, hope to be able to contribute a little share to the further proliferation and implementation of LCC ideas and practices, together with other ongoing activities as the new SETAC Working Group on life cycle costing … " (Hunkeler and Rebitzer 2003; Rebitzer and Seuring 2003).

LCC is one pillar of sustainability. Sustainability comprises three pillars: environment, economy and social aspects (Rebitzer and Hunkeler 2003). "For the environmental part there is already an internationally standardized tool: Life Cycle Assessment (LCA). Life Cycle Costing (LCC) is the logical counterpart of LCA for the economic assessment. LCC surpasses the purely economic cost calculation by taking into account the use- and end-of-life phases and hidden costs" (Klöpffer 2003, 2008).

In 2011 SETAC published a code of practice for environmental life cycle costing (LCC).

"The objective of the code of practice is to provide readers with a solid understanding of how to apply LCC in parallel with LCA to stimulate additional case studies and peer-reviewed research to further refine the methodology. The ultimate goal is to build consensus for an international standard that parallels the ISO 14040 standard for LCA" (Swarr et al. 2011).

Life-cycle costing: a Code of Practice (98 pp), is published by SETAC Press and is available at https://www.setac.net/setacssa/ecssashop.show_product_detail?p_mode=detail&p_product_serno=374 for $5 for members and $12 for non-members. It is based on the deliberations of the SETAC Working Group on Life-Cycle Costing.

Being closely connected with sustainability, LCC has attracted much interest in Int J Life Cycle Assess between 2003 and 2012.

6.3 *Social Life Cycle Assessment (SLCA)*

"Social LCA aims at facilitating companies to conduct business in a socially responsible manner by providing information about the potential social impacts on people caused by the activities in the life cycle of their product" (Dreyer et al. 2006).

"Social life cycle assessment (S-LCA) emerged in the last years as a methodological approach aimed at evaluating social and socioeconomic aspects of products and their potential positive and negative impacts along their life cycle. According to the Guidelines for social life cycle assessment of products (Benoît and Mazijn 2009), developed within the UNEP/SETAC Life Cycle Initiative, social impacts are those that may affect stakeholders along the life cycle of a product and may be linked to company behaviour, socioeconomic processes and impacts on social capital" (Zamagni et al. 2011).

As early as 1996, O'Brien et al. discussed the emerging theme 'Social Life Cycle Assessment' in Int J Life Cycle Assess They combined environmental life cycle assessment (ELCA) and social life cycle assessment (SLCA) and called this approach Social and Environmental Life Cycle Assessment (SELCA). "The value of the approach lies in establishing what social action, as well as what technical developments, may be undertaken in order to effect positive change within the industrial or commercial cycle under investigation" (O'Brien et al. 1996).

Since 2005 SLCA has been developing, which is reflected in Int J Life Cycle Assess "It is clear that the assessment of the social aspects of all elements of the life cycle is a critical future issue for life cycle approaches in general" (Hunkeler and Rebitzer 2005).

In 2006 David Hunkeler integrated the subject area 'Societal LCA' into the journal. "Int J LCA clearly has a lead role in the development and proliferation of LCA thinking and applications and, as such, it can become the vehicle for LCA-compatible societal assessments … " (Hunkeler 2006a).

Hunkeler defines Social Assessment as likely to be based on mid-point indicators, whereas Societal Assessment might be more macroeconomic and hence end-point based.

The goal of societal life cycle assessment is not to make decisions, but to point out tradeoffs to decision- or policy-makers (Hunkeler 2006b).

In 2004, UNEP/SETAC Life Cycle Initiative recognised the need for a task force on the integration of social criteria into LCA. The publication of the **Guidelines for Social Life Cycle Assessment**[6] **of Products** (eds Benoit and Mazijn 2009) was launched officially on 18th May, 2009, in Quebec, Canada.

These Guidelines ground the assessment of the social and socio-economic aspects into the LCA framework. The proposed framework is in line with the ISO 14040 and 14044 LCA standards but adapted for the social aspects.

The Guidelines contain four main sections:

1. The first section presents the historical context in which the guidelines should be placed. From the broad and vague concept of sustainable development to the more specific goal of sustainable consumption and production.
2. The second section explains the principles of environmental life cycle assessment and life cycle costing
3. The third section provides a technical framework for SLCA. The four major phases (goal and scope of the study, inventory analysis, impact assessment, and interpretation as outlined in ISO 14040 and 14044) of the methodology are presented.

[6] In this context, Social LCA means social and socio-economic LCA.

4. The fourth section presents the possible applications and the limitations, the communication of results, the review process, and identified research and development needs (Benoît et al. 2010).

The 2nd International Seminar in Social Life Cycle Assessment was held on 5 and 6 May 2010 in Montpellier. It was a follow-up to the first seminar held in Lyngby at the Denmark Technology University on 31 May 2010, initiated by Louise Camilla Dreyer.

The variety of speeches highlighted the different methodologies in social LCA which concurrently have emerged (Macombe et al. 2011)

- Management SLCA is devoted to internal decision making within a value chain and to the identification of social hot spots.
- Consequential LCA aims to assess the social impacts caused by choosing between decision alternatives.
- Educative SLCA communicates the preference of the decision maker to the market.

6.4 *Life Cycle Sustainability Assessment (LCSA)*

"Sustainable development is development that meets the needs of present without compromising the ability of future generations to meet their own needs" (Brundtland Report[7]).

"The concept of 'sustainable development' (or 'sustainability') was introduced by the United Nations in the Brundtland declaration but has eluded precise definition. In very broad terms, sustainable development means a pattern of human activity that is consistent with the ecological and thermodynamic maintenance of the planet, which is technically and economically viable, and which meets people's needs and expectations .... The idea is summed up in Fig. 5.1. 'Sustainable Development' is the area at the centre of the diagram where the 'natural', 'techno-economic' and 'social' intersect" (O'Brien et al. 1996).

Figure 5.1 in the article by O'Brien, Alison Doig and Roland Clift (1996) is a convincing illustration of 'Sustainable development' in Int J Life Cycle Assess (Fig. 5.4).

"Sustainability—a term originating from silviculture, which was adopted by UNEP as the main political goal for the future development of humankind—is also the ultimate aim of product development. It comprises three components: environment, economy and social aspects which have to be properly assessed and balanced if a new product is to be designed or an existing one is to be improved" (Klöpffer 2008).

[7] The Brundtland Report of 1987 is also known as Our Common Future. Formally known as the 'World Commission on Environment and Development' (WCED), the Brundtland Commission's mission was to unite countries to pursue sustainable development. The Chairman of the Commission was Gro Harlem Brundtland, Norway.

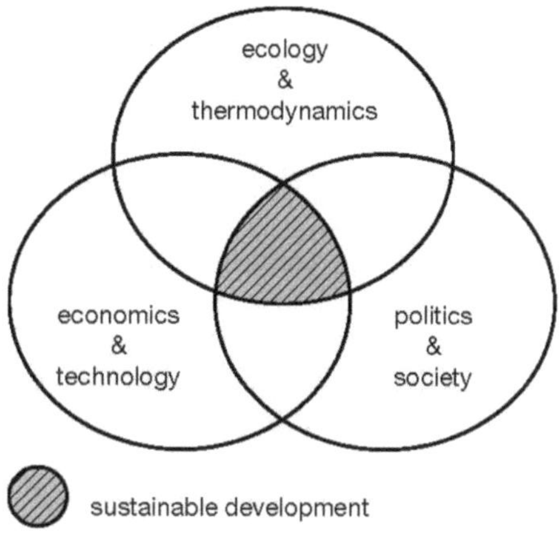

Fig. 5.4 Components of sustainable development

There is not much difference between the two definitions of 'Sustainability' by O'Brien et al. in 1996 and Klöpffer in 2008. They ground on the 'three pillar equation' or 'triple bottom line'. This interpretation means that, for achieving and assessing sustainability, the environmental (LCA), economic (LCC) and social (SLCA) aspects have to be integrated.

The idea of combining three LCA techniques (methods) into an LCSA was first formulated by Walter Klöpffer (2008), followed by Matthias Finkbeiner et al. (2010).

$$\text{LCSA} = \text{LCA} + \text{LCC} + \text{SLCA}$$

LCSA: Life Cycle Sustainability Assessme
LCA: (environmental) Life Cycle Assessment
LCC: (environmental) Life Cycle Costing
SLCA: Social Life Cycle Assessment

Valdivia et al. (2012) identified that, while LCSA is feasible, the following areas need more development: data production and acquisition, methodological development, discussion about LCSA criteria (e.g. cutoff rules), definitions and formats of communication and dissemination of LCSA results and the expansion of research and applications combining (environmental) LCA, LCC and SLCA.

Alessandra Zamagni, the subject editor for LCSA in Int J Life Cycle Assess, invited practitioners and method developers to submit articles addressing the full range of sustainability-related topics, also case studies, methodological developments, discussions about data availability, and thus how the present software tools can deal with such evaluation are encouraged (Zamagni 2012; Zamagni et al. 2013).

From the preface:

"SLCA has been neglected in the past, but is now beginning to be developed. One of the challenges is how to relate the social indicators (social impact assessment) to the functional unit of the product-system and how to restrict the many social indicators proposed to a manageable number. Meanwhile, qualitative and semi-quantitative approaches are used as substitutes for a full, quantitative SLCA. It is hoped that new methods will be developed and finally standardised by ISO. The combination of LCA, LCC and SLCA, represented by the three-pillar equation LCSA=LCA+LCC+SLCA, will provide the much needed tool for sustainability assessment of products.

However, broadening indicators is not enough, since it is also necessary to further sophisticate and deepen the modelling, in order to address complexities and sustainability questions along the full range of scales (from local to global), taking more mechanisms and relations into account. Mechanisms are connecting links between activities and they can show up everywhere, involving a variety of domains and giving rise to different consequences. Market mechanisms are part of broader economic mechanisms, which recall concepts like employment and growth. These in turn function within a cultural, social, political and regulatory context. All of this could be achieved through the development of new approaches or through the combination or integration of LCA with other methods, while managing or counteracting the resulting increase in complexity".

The following questions, raised by Alessandra Zamagni (2012), still need to be answered:

- How can the LCSA framework be consistently applied, considering also the different degree of maturity of the three methods?
- What role does scenario modelling play in the LCSA framework?
- What other approaches to LCSA can be proposed than the three separate assessments?
- What approaches exist for including mechanisms in the analysis? How can different domains, normative positions (values) and empirical knowledge be dealt with? How can future changing structures of the economy be accounted for? And what kind of methods and tools can be used, combined and/or integrated?
- What do we need to further develop LCSA? What research strategies and lines are considered relevant?
- How can uncertainty, which is an inevitable and inherent characteristic of sustainability assessment, be accommodated and managed?

7 Special Issues and Supplements

Int Journal of Life Cycle Assess distinguishes between special issues and supplemental issues, although both were generally referred to as 'special issues' until recently[8]. Both are edited by invited guest editors. Special issues belong to the regular series of printed issues, with the title and the names of the guest editors printed on the cover. Supplements appear outside of the regular series and are paid by the commissioner; articles have to be cited as "Author(s) (year) Title of the paper Int J Life Cycle Assess (Vol No) Supplement No, first page—last page."

[8] The distinction between 'special issue' and 'supplemental issue' was enacted with the transfer of the journal to Springer-Verlag in 2008.

The first Special Issue in Int J Life Cycle Assess was published in 1996 and 1997, in the first and second volume of the journal. It was called 'Taormina Issue' and consisted of 13 selected papers from the 6th SETAC Europe Meeting on May 19–22, 1996 in Taormina, Sicily. This Taormina issue created the beginning of an ongoing publication of special issues and supplements in the journal.

Special issues include the following topics and editors:

- Selected Papers from the 6th SETAC Europe Meeting, May 19–22, 1996, Taormina, Sicily (vol. 1 and 2, 1996). Special Edition Editor: Allan Astrup Jensen; co-editors: Roland Clift, Patrick Hofstetter and Dennis Postlethwaite
- The MIIM LCA Ph.D. Club. Special issue vol. 4, 1999, vol. 5, 2000
- LCA in Japan. Special issue vol. 5, no. 5, 2000. Special Edition Editors: Matthias Finkbeiner and Yasunari Matsuno
- The International Conferences on Life Cycle Assessment
  - The International Conference on Life Cycle Assessment, Arlington, Virginia, USA, 2000. Special issue vol. 6, no. 2, 2001. Special Edition Editors: Mary Ann Curran and Rita Schenck
  - The International Conference Life Cycle Assessment/Life Cycle Management: A Bridge to a Sustainable Future, Seattle, Washington, USA, 2003. Special issue vol. 9, no. 6, 2004 and vol. 10, 2005. Special Edition Editor: Mary Ann Curran
- OMNIITOX (Operational Models aNd Information tools for Industrial applications of eco/TOXicological impact assessments). Special issue vol. 9, no. 5, 2004. Special Edition Editor: David W. Pennington.
- The ecoinvent database. Special issue vol. 10, no. 1, 2005. Special Edition Editor: Rolf Frischknecht
- 'Sustainable Management of Natural Resources in an Life-cycle Perspective'. Special Issue vol. 11, no. 1, 2006, inspired by the SETAC World Conference in Portland (USA), November 2004 (in co-operation with the UNEP/SETAC Life Cycle Initiative). Special Edition Editor: Helias A Udo de Haes
- Honouring Helias Udo de Haes. Special issue vol. 11, no. 1, January 2006. Special Edition Editors: Mark AJ Huijbregts, Jeroen B Guinée, Gjalt Huppes, José Potting
- LCM 2007 Zurich—From Analysis to Implementation. 3rd International Conference on Life Cycle Management, Zurich, August 27–29, 2007. Special issue vol. 12, no. 1, August 2007. Special Edition Editors: Gerald Rebitzer, Stefanie Hellweg, Annette Koehler
- Life Cycle Performance of Aluminium Applications. Supplement vol. 14, no. 1, May 2009. Special Edition Editors: Gerald Rebitzer, Jörg H Schäfer
- LCIA of impacts on human health and ecosystems (USEtox). Special issue vol. 16, no. 8, September 2011. Special Edition Editors: Michael Z Hauschild, Olivier Jolliet, Mark AJ Huijbregts
- Promotion of Young Scientists in LCA. Special issue vol. 17, no. 9, November 2012. Special Edition Editor: Liselotte Schebek

- Global Land Use Impacts on Biodiversity and Ecosystem Services in LCA. Special issue vol. 18, no. 6, July 2013. Special Edition Editors: Thomas Koellner, Roland Geyer
- Life Cycle Sustainability Assessment: From LCA to LCSA. Special issue vol. 18, no. 9, November 2013. Special Edition Editors: Alessandra Zamagni, Hanna-Leena Pesonen, Thomas Swarr

8 ISO Standardisation of LCA

International standards for LCA were developed since the 1990s by ISO Technical Committee (TC) 207 (Environmental Management) as part of the ISO 14,000 family of environmental management standards. The committee within ISO/TC207 dealing with LCA is Subcommittee 5 (SC5). So the complete name of the LCA unit is ISO/TC207/SC5.

A comprehensive coverage of the history, present, and future of the ISO standardisation of LCA is given by Matthias Finkbeiner in Chapter 3, this volume[9] (Finkbeiner 2013).

"International standards on Life Cycle Assessment are only significant if they make the necessary practical instructions without extending into regulations which may be far too detailed. In other words, a balance must be achieved between the unavoidable establishment and the possibility of interpreting these regulations more or less freely" (Marsmann 2000).

The articles on ISO-LCA in Int J Life Cycle Assess (Table 5.1) reflect the development of the standardisation process.

This division of LCA methodology into successive phases was directly inspired by the SETAC 'Code of Practice'(1991) which was the most authoritative publication to be referred to.

ISO 14040 (1997) = Principles and framework
ISO 14041 (1998) = Goal and scope definition, inventory
ISO 14042 (2000) = Impact assessment
ISO 14043 (2000) = Interpretation (formerly 'Improvement')
ISO 14040 and ISO 14044 (2006) have become the commonly accepted rules for LCA. They are the 'core standards':
ISO 14040: Environmental Management—Life Cycle Assessment—Principles and Framework
ISO 14044: Environmental Management—Life Cycle Assessment—Requirements and Guidelines

[9] The international standards as the constitution of LCA: the ISO 14040 series and its offspring by Matthias Finkbeiner.

Table 5.1 Articles on LCA ISO standardisation in Int J Life Cycle Assess

ISO 14040	Environmental Management – Life Cycle Assessment – Principles and Framework (1997 and 2006)	Int J Life Cycle Assess (1997) 2(3): 121 ISO 14040 Angela Merkel Int J Life Cycle Assess (1997) 2(3): 122–123 ISO 14040 – The First Project Manfred Marsmann "The introduction of ISO 14040 sets a process in motion which is now unstoppable. Within a short period of time we shall have four standards which combine the elements of life cycle assessment, as far as this is possible, in a way which is comprehensive yet practical, standardised yet flexible and precise yet comprehensible." Int J Life Cycle Assess (1997) 2(4): 183–184 Peer (Expert) Review in LCA According to SETAC and ISO 14040 – Theory and Practice Walter Klöpffer	Int J Life Cycle Assess (1997) 2(1): 2–4 **Special Issue: Current LCA-ISO Activities** Foreword – Development of Life Cycle Thinking – ISO Standards – Standardization of – Environmental Balances: ISO 14040 – Subsequent Standards – Inventory: ISO 14 041 – Life Cycle Impact Assessment: ISO 14 042 – Interpretation of Results: ISO 14043 Manfred Marsmann, Hans-Jürgen Klüppel, Konrad Saur Int J Life Cycle Assess (1997) 2(2): 64–65 **Special Issue: Current LCA-ISO Activities** Brief Result Report of WGs of SC 5 'Life Cycle Assessment' on the Kyoto Meeting of the ISO/TC 207 Gertraud Goldhan, Sabine Schlüter
ISO 14041	Environmental Management – Life Cycle Assessment – Goal and Scope Definition and Inventory Analysis (1998)	Int J Life Cycle Assess (1997) 2(1): 5–8 Goal and Scope Definition and Life Cycle Inventory Analysis Hans-Jürgen Klüppel Int J Life Cycle Assess (1998) 3(6): 301 Goal and Scope Definition and Life Cycle Inventory Analysis Hans-Jürgen Klüppel	Int J Life Cycle Assess (2000) 5(6): 317–318 The ISO 14040 Family Manfred Marsmann
ISO 14042	Environmental Management – Life Cycle Assessment – Life Cycle Impact Assessment (2000)	Int J Life Cycle Assess (1997) 2(2): 66–70 Life Cycle Impact Assessment Konrad Saur Int J Life Cycle Assess (1998) 3(4): 180–181 Letter to the Editor ISO 14042 Restricts Use and Development of Impact Assessment Commentary by Edgar G. Hertwich and William S. Pease Int J Life Cycle Assess (1999) 4(2): 65 Letter to the Editor In Reply to Hertwich & Pease,	Int J Life Cycle Assess (2002) 7(1): 1 The ISO Standardization Process: Quo Vadis? Hans-Jürgen Klüppel Int J Life Cycle Assess (2005)10(3): 165 The Revision of ISO Standards 14040–14043 ISO 14040: Environmental management – Life cycle assessment – Principles and framework ISO 14044: Environmental management – Life cycle

Table 5.1 (continued)

		Int. J. LCA 3 (4) 180–181, 'ISO 14042 Restricts Use and Development of Impact Assessment' Manfred Marsmann, Sven Olaf Ryding, Helias Udo de Haes, James Fava, Willie Owens, Kevin Brady, Konrad Saur, Rita Schenck Int J Life Cycle Assess (1999) 4(2): 75–80 Commentary Article How Does ISO/DIS 14042 on Life Cycle Impact Assessment Accommodate Current Best Available Practice? Helias A. Udo de Haes, Olivier Jolliet Int J Life Cycle Assess (1999) 4(6): 307 ISO 14042 Sven-Olof Ryding	assessment – Requirements and guidelines Hans-Jürgen Klüppel Int J Life Cycle Assess (2006) 11(2): 80–85 The New International Standards for Life Cycle Assessment: ISO 14040 and ISO 14044 Matthias Finkbeiner, Atsushi Inaba, Reginald B.H. Tan, Kim Christiansen, Hans-Jürgen Klüppel Int J Life Cycle Assess (2012) 17(9): 1087–1093 The critical review of life cycle assessment studies according to ISO 14040 and 14044: Origin, purpose and practical performance Walter Klöpffer
ISO 14043	Environmental Management – Life Cycle Assessment – Life Cycle Interpretation (2000)	Int J Life Cycle Assess (1997) 2(1): 8–10 Life Cycle Interpretation – A Brand New Perspective? Konrad Saur "In my personal opinion, the interpretation step is the key element toward reliability and an acceptance of the whole LCA framework." Int J Life Cycle Assess (1999) 4(5) 245 ISO 14043 – Life Cycle Interpretation Henri Lecouls	Int J Life Cycle Assess (2013) 18(2):300–301 Letter to the Editor-in-Chief: Regarding your article 'The critical review of life cycle assessment studies according to ISO 14040 and 14044—origin, purpose and practical performance', Int J Life Cycle Assess (2012) 17: 1087–1093. Christoph Koffler
ISO 14044	Environmental Management – Life Cycle Assessment – Requirements and Guidelines (2006)	Int J Life Cycle Assess (2005) 10(6): 381 Letter to the Editor ISO 14044 also Applies to Social LCA Bo Weidema	Int J Life Cycle Assess (2013) 18(1): 1–4 From the 40s to the 70s—the future of LCA in the ISO 14000 family Matthias Finkbeiner

9 Conclusion

The International Journal of Life Cycle Assessment is still the only scientific journal devoted entirely to LCA methods and LCM application. Over the years, its scope broadened with the development of life-cycle based methods exceeding the classical LCA, as defined by SETAC (1993) and ISO (1997 ff.).

From the beginning, the journal has been a harbour for LCA societies around the world. In 2003, the UNEP/SETAC Life Cycle Initiative established an official collaboration with the Journal, making it the Associated Journal of the UNEP/SETAC

Life Cycle Initiative. Although some societies ceased their regular documentation after a few years, societies have been important contributors of editorials and scientific papers. This is evidence that the association with Int J Life Cycle Assess has been functional and successful.

The basic editing philosophy, namely to publish method developments as well as applied papers, has not been changed since the first issue. Each article has to present new information or data, such as previously undisclosed foreground data, or advance understanding and knowledge in the field. As a new variant is identified as sufficiently matured (at least to a certain degree), motivated editors are invited to develop the special field.

Furthermore contributions from remote areas of the world are very welcomed. The journal editors continue to strive to maintain truly global authorship and readership while ensuring confidence in the scientific level and the practical usefulness of the journal's contributions to the open literature.

References

Baitz M, Albrecht S, Brauner E, Broadbent C, Castellan G, Conrath P, Fava J, Finkbeiner M, Fischer M, Fullana i Palmer P, Krinke S, Leroy C, Loebel O, McKeown P, Mersiowsky I, Möginger B, Pfaadt M, Rebitzer G, Rother E, Ruhland K, Schanssema A, Tikana L (2013) Editorial. LCA's theory and practice—like ebony and ivory living in perfect harmony? Int J Life Cycle Assess 18(1):5–13

Baldo GL, Giacomucci A (1999) The LCA society of Italy. Int J Life Cycle Assess 4(4):228

Baldo GL, Pretato U (2001) A database to support Italian LCA practitioners. Int J Life Cycle Assess 6(3):180

Baumann H (2002) Letter to the editor: Int J LCA could have received better acknowledgement. Int J Life Cycle Assess 7(1):2–3

BenoÈt C, Mazijn B (2009) (eds) Guidelines for social life cycle assessment of products. Paris. http://lcinitiative.unep

Benoît C, Norris GA, Valdivia S, Ciroth A, Moberg A, Bos U, Prakash S, Ugaya C, Beck T (2010) The guidelines for social life cycle assessment of products: just in time! Int J Life Cycle Assess 15:156–163

Boura A, Koroneos CJ, Moussiopoulos N (2000) Overview of LCA activities in Greece from LHTEE. Int J Life Cycle Assess 5(4):203–204

Braunschweig A (2005) Quality control and peer review. The 23rd LCA discussion forum at ETH Zürich (23rd September 2004). Int J Life Cycle Assess 10(2):158

Bretz R (1998) SPOLD (editorial). Int J Life Cycle Assess 3(3):119–120

Cappellaro F, Scalbi S, Masoni P (2008) The Italian network on LCA (editorial). Int J Life Cycle Assess 13(7):523–525

Curran MA (2006) Report on activity of task force 1 in the life cycle inventory programme: Data registry—global life cycle inventory data resources. Int J Life Cycle Assess 11(4):284–289

Datta A, Valdivia S (2013) Indian life cycle assessment and management conference (ILCM) 2013 creating business value through sustainable strategies. Int J Life Cycle Assess 18(5):1173

Datta A, Strothmann P, Valdivia S, Vigon B (2012) First Indian life cycle assessment and management conference 2012. Int J Life Cycle Assess 17(7):944–945

Domenéch X, Fullana P (1996) LCA association in Spain. Int J Life Cycle Assess 1(4):207

Doublet G, Jungbluth N (2011) Environmental product information (EPI) and LCA, Ittingen-Berne, Switzerland, June 22, 2010. 41st LCA Swiss discussion forum. Int J Life Cycle Assess 16:90–94

Dreyer L, Hauschild M, Schierbeck J (2006) A framework for social life cycle impact assessment. Int J Life Cycle Assess 11(2):88–97
Fava J, Sonnemann G, Gloria T (2006) Key observations arising from papers on sustainable production, use and recycling of natural resources. Int J Life Cycle Assess 11(1):75–76
Finkbeiner M (2013) From the 40s to the 70s—the future of LCA in the ISO 14000 family (editorial). Int J Life Cycle Assess 18:1–4
Foley J, Lant P (2009) Regional normalisation figures for Australia 2005/2006—inventory and characterisation data from a production perspective. Int J Life Cycle Assess 14:215–224
Friot D, Loerincik Y, Jolliet O (2005) Life cycle approaches for green investment. 26th LCA Swiss discussion forum. Int J Life Cycle Assess 10(6):454–456
Frischknecht R (2000) How or to whom do we communicate LCA results? Comments to the editorial in Int. J. LCA 5 (3) 125 (2000) by Rolf Frischknecht. Int J Life Cycle Assess 5(4):187
Frischknecht R, Flury K (2011) Life cycle assessment of electric mobility: answers and challenges—Zurich, April 6, 2011. 43rd LCA Swiss discussion forum. Int J Life Cycle Assess 16:691–695
Frischknecht R, Büsser S, Krewitt W (2009) Environmental assessment of future technologies: how to trim LCA to fit this goal? 38th LCA Swiss discussion forum. Int J Life Cycle Assess 14:584–588
Giacomucci A, Baldo GL (1998) LCA activities in Italy. Int J Life Cycle Assess 3(2):68
Giacomucci A, Baldo GL (1999) LCA certification according to ISO 14.040—first experience. Int J Life Cycle Assess 4(3):165–166
Grant T (2000) Moving from problems to solutions—2nd national LCA conference, Melbourne, Australia 23rd to 24th February 2000. Int J Life Cycle Assess 5(2):74
Grant T (2002) The third Australian conference on life cycle assessment (conference report). Int J Life Cycle Assess 7(5):308
Grant T, James K (2002) Australian LCA society (ALCAS). Int J Life Cycle Assess 7(3):177
Grant T, James KL, Lundie S (2001) Australian LCA society (editorial). New LCA group to promote life cycle assessment in Australia. Int J Life Cycle Assess 6(6):323–324
Hauschild M (1998) Book review. Analisis del ciclo de vida—life cycle assessment in Spanish. Authors: Pere Fullana, Rita Puig. Int J Life Cycle Assess 3(1):56
Hauschild MZ, Jolliet O, Huijbregts MA (2011) LCIA of impacts on human health and ecosystems (USEtox). A bright future for addressing chemical emissions in life cycle assessment (editorial). Int J Life Cycle Assess 16(8):697–700
Heinrich AB, Klöpffer W (1999) Online-first. Int J Life Cycle Assess 4(6):340
Heinrich AB, Klöpffer W (2002) LCM—integrating a new section in Int J LCA (editorial). Int J Life Cycle Assess 7(6):315–316
Hindle P, De Oude NT (1996) SPOLD—society for the promotion of life cycle development. Int J Life Cycle Assess 1(1):55–56
Hischier R, Baitz M, Bretz R, Frischknecht R, Jungbluth N, Marheineke sT, McKeown P, Oele M, Osset P, Renner I, Skone T, Wessman H, De Beaufort ASH (2001) SETAC Europe LCA working group 'Data Availability and Data Quality'. Guidelines for consistent reporting of exchanges from/to nature within life cycle inventories (LCI). Int J Life Cycle Assess 6(4):192–198
Hofstetter P (2000) How or to whom do we communicate LCA results? Comments to the editorial in Int. J. LCA 5 (3) 125 (2000) by Patrick Hofstetter. Int J Life Cycle Assess 5(4):187
Hunkeler D (2006a) Societal assessment—a new subject area and call for papers. Int J Life Cycle Assess 11(2):86
Hunkeler D (2006b) Societal LCA methodology and case study. Int J Life Cycle Assess 11(6):371–382
Hunkeler D, Rebitzer G (2003) Life cycle costing—paving the road to sustainable development? Int J Life Cycle Assess 8(2):109–110
Hunkeler D, Rebitzer G (2005) The future of life cycle assessment. Int J Life Cycle Assess 10(5):305–308
Hunkeler D, Yamamoto R, Yasui I (1999) Life cycle assessment: from Adam Smith to Aristotle. A review of the Third International Conference on Ecobalances. Tsukuba, Japan, November 25–27, 1998

Hunkeler D, Rebitzer G, Jensen AA, Margni M (2001) Life cycle management—bridging the gap between science and application. Report from LCM 2001—First international conference on life cycle management, Copenhagen, August 27–29, 2001. Int J Life Cycle Assess 6(6):384–390

Hunkeler D, Suh S, Finnveden G, Fava J, Udo deHHA, Baitz M, Finkbeiner M, Young A, Heinrich AB (2008) Dedicated to Walter Klöpffer on the occasion of his anniversary on 6 June 2008. Int J Life Cycle Assess 13(3):177

Itsubo N (1999a) The progress of the impact assessment study committee in the national LCA project of Japan. Int J Life Cycle Assess 4(4):194

Itsubo N (1999b) The progress of inventory study committee WG2 in the national LCA project in Japan. Int J Life Cycle Assess 4(5):246

Itsubo N (2000) Launch of the damage function sub-committee in the national LCA project of Japan. Int J Life Cycle Assess 5(2):84

James K (2003) Water issues in Australia—an LCA perspective. Summary of presentations made at the ALCAS LCA Roundtable held Wednesday 7th May 2003. Int J Life Cycle Assess 8(4):242

James K, Grant T (2002) What's happening in Australia? Int J Life Cycle Assess 7(1):4

James K, Narayanaswamy V (2004) ALCAS: best practice seminar and announcement of forthcoming 2005 conference. Int J Life Cycle Assess 9(4):214–215

James K, Grant T, Sonneveld K (2002) Stakeholder involvement in Australian paper and packaging waste management LCA study. Int J Life Cycle Assess 7(3):151–157

James K, Beavis P, Grant T (2003) ALCAS: reflections upon 2002 and looking ahead to 2003. Int J Life Cycle Assess 8(1):49

Jensen AA, Hunkeler D, Gaillard G, Hellweg S, Christiansen K (2005) LCM 2005—innovation by life cycle management. Barcelona, 5–7 September 2005. Int J Life Cycle Assess 10(6):451–453

Jolliet O, Müller-Wenk R, Bare J, Brent A, Goedkoop M, Heijungs R, Itsubo N, Peñia C, Pennington D, Potting J, Rebitzer G, Stewart M, Udo deHH, Weidema B (2004) The LCIA midpoint-damage framework of the UNEP/SETAC Life Cycle Initiative. Int J Life Cycle Assess 9(6):394–404

Jolliet O, Dubreuil A, Gloria T, Hauschild M (2005) Progresses in life cycle impact assessment within the UNEP/SETAC Life Cycle Initiative. Int J Life Cycle Assess 10(6):447–448

Kim I, Hur T (2009) Integration of working environment into life cycle assessment framework. Int J Life Cycle Assess 14:290–301

Klöpffer W (1996) Editorial. Int J Life Cycle Assess 1(1):3

Klöpffer W (1997) ISOLP—do we truly require an international society for LCA practitioners (editorial)? Int J Life Cycle Assess 2(1):1

Klöpffer W (2003) Life-cycle based methods for sustainable product development. Int J Life Cycle Assess 8(3):157–159

Klöpffer W (2004) Analytical tools for environmental design and management in a systems perspective. The combined use of analytical tools (book review). Int J Life Cycle Assess 9(2):137

Klöpffer W (2005) The critical review process according to ISO 14040–43. An analysis of the standards and experiences gained in their application. Int J Life Cycle Assess 10(2):98–102

Klöpffer W (2007) Publishing scientific articles with special reference to LCA and related topics. Int J Life Cycle Assess 12(2):71–76

Klöpffer W (2008) State-of-the-art in life cycle sustainability assessment (LCSA). Life cycle sustainability assessment of products (with comments by Helias A. Udo de Haes, p. 95). Int J Life Cycle Assess 13(2):89–95

Klöpffer W (2012) The critical review of life cycle assessment studies according to ISO 14040 and 14044—origin, purpose and practical performance. Int J Life Cycle Assess 17(9):1087–1093

Klöpffer W (2013) Comment on the editorial note by Baitz et XXI aliis. Int J Life Cycle Assess 18(1):14–16

Klöpffer W, Heinrich AB (1999) Can you imagine the LCA world without our journal (editorial)? Int J Life Cycle Assess 4(4):183

Klöpffer W, Heinrich AB (2001) Two planets and one journal (editorial). Int J Life Cycle Assess 6(1):1–3

Klöpffer W, Heinrich AB (2002) Life cycle management (LCM) (editorial). Int J Life Cycle Assess 7(3):133
Klöpffer W, Heinrich AB (2009) How to communicate LCA results (editorial). Int J Life Cycle Assess 5(3):125
Loerincik Y, Kaenzig J, Jolliet O (2005) Life cycle approaches for sustainable consumption. 24th LCA Swiss discussion forum. Int J Life Cycle Assess 10(3):228–229
Macombe C, Feschet P, Garrabé M, Loeillet D (2011) 2nd international seminar in social life cycle assessment—recent developments in assessing the social impacts of product life cycles. Int J Life Cycle Assess 16:940–943
Marsmann M (2000) The ISO 14040 family (editorial). Int J Life Cycle Assess 5(6):317–318
Matsuno Y, Kondo Y (2007a) Corner: J LCA Jpn (The Journal of Life Cycle Assessment Japan) Eds Yasunari Matsuno Yasushi Kondo Announcing the collaboration between Int J LCA J LCA Jpn. Int J Life Cycle Assess 12(6):348–350
Matsuno Y, Kondo Y (2007b) Corner: J LCA Jpn (The Journal of Life Cycle Assessment Japan) Eds Yasunari Matsuno Yasushi Kondo The collaboration between Int J LCA J LCA Jpn. Int J Life Cycle Assess 12(7):547–549
Matsuno Y, Kondo Y (2008a) Corner: J LCA Jpn (The Journal of Life Cycle Assessment, Japan). Eds Yasunari Matsuno and Yasushi Kondo. The collaboration between Int J LCA and J LCA Jpn. Int J Life Cycle Assess 13(1):10–11, 13(2):168–171
Matsuno Y, Kondo Y (2008b) The collaboration between Int J Life Cycle Assess and J LCA Jpn. Int J Life Cycle Assess 13(7):605–608
Matsuno Y, Itsubo N, Hondo H, Nakano K (2013) LCA in Japan in the twenty-first century. Int J Life Cycle Assess 18(1):278–284
May JR, Brennan DJ (2003) Application of data quality assessment methods to an LCA of electricity generation. Int J Life Cycle Assess 8(4):215–225
Morimoto M (1997) JLCA Corner: activities of the life-cycle assessment society of Japan (JLCA). Int J Life Cycle Assess 2(3):153
Moussiopoulos N, Koroneos CJ (1998a) First LCA workshop in Greece. Int J Life Cycle Assess 3(5):288
Moussiopoulos N, Koroneos CJ (1998b) Hellenic life cycle assessment network (HELCANET). Int J Life Cycle Assess 3(5):272
Nakahara Y (1999) The progress of the database study committee in the national LCA project of Japan. Int J Life Cycle Assess 4(6):310
Nakano K, Nakaniwa C, Kabeya T, Iguchi T, Aoki R (2007) Current activities of the life cycle assessment society of Japan. Int J Life Cycle Assess 12(7):546
Nebel B (2006) Second LCA workshop/roundtable, Rotorua, Scion, NZ, February 2006. Int J Life Cycle Assess 11(4):292–293
Nebel B, Nielsen PS (2005) LCA Workshop/roundtable, Rotorua, Forest Research, NZ, February 2005. Int J Life Cycle Assess 10(5):375–376
O'Brien M, Doig A, Clift R (1996) 6th SETAC Europe meeting: LCA—selected papers. Social and environmental life cycle assessment (SELCA). Approach and methodological development. Int J Life Cycle Assess 1(4):231–237
Owens W (2000) How or to whom do we communicate LCA results? Comments to the Editorial in Int. J. LCA 5 (3) 125 (2000) by William Owens. Int J Life Cycle Assess 5(4):188
Parsons D (2007) The environmental impact of disposable versus re-chargeable batteries for consumer use. Int J Life Cycle Assess 12(3):197–203
Parsons D (2010) Home heating in temperate Australia. Int J Life Cycle Assess 15:785–793
Peters GM, Wiedemann SG, Rowley HV, Tucker RW (2010) Accounting for water use in Australian red meat production. Int J Life Cycle Assess 15:311–320
Potting J, Curran MA, von Blottnitz H (2010) From life cycle talking to taking action. The 4th international conference on life cycle management—the global challenge of managing life cycles. Int J Life Cycle Assess 15:326–329
Puri P, Compston P, Pantano V (2009) Life cycle assessment of Australian automotive door skins. Int J Life Cycle Assess 14:420–428

Rebitzer G, Hunkeler D (2003) Life cycle costing in LCM—ambitions, opportunities, and limitations (editorial). Discussing a framework. Int J Life Cycle Assess 8(5):253–256 (2003
Rebitzer G, Seuring S (2003) Life cycle costing—a new SETAC Europe working group. Methodology and application of life cycle costing. Int J Life Cycle Assess 8(2):110–111
Roos P (2007) Be welcome at Springer, unit 'Environmental Sciences'. Int J Life Cycle Assess 12(7):347
Saner D, Walser T, Vadenbo CO (2012) End-of-life and waste management in life cycle assessment—Zurich, 6 December 2011. 46th LCA Swiss discussion forum. Int J Life Cycle Assess 17:504–510
Schuerch R, Kaenzig J, Jungbluth N, Nathani C (2012) 45th Discussion forum on LCA—environmentally extended input–output analysis and LCA, September 15, 2011, Berne, Switzerland. Int J Life Cycle Assess 17:840–844
SETAC—Society of Environmental Toxicology and Chemistry (1993) Guidelines for life-cycle assessment: a 'Code of Practice'. From the SETAC Workshop held at Sesimbra, Portugal, 31 March–3 April 1993. Edn. 1, Brussels and Pensacola (Florida), August 1993
Sharma V (1999) National Ecology and Environment Foundation (NEEF). Int J Life Cycle Assess 4(6):356
Sharma V (2000) Indian Society for Life Cycle Assessment (ISLCA). Int J Life Cycle Assess 5(1):4
Siegenthaler CP, Margni M (2005) Dissemination, application and assessment of LCA in industry. 25th LCA discussion forum at ETH Zürich, Switzerland. Int J Life Cycle Assess 10(5):377–378
Singhofen A, Hemming CR, Weidema BP, Grisel L, Bretz R, De Smet B, Russell D (1996) Life cycle inventory data—development of a common format. Int J Life Cycle Assess 1(3):171–178
Sonnemann G (2005) International conference on life cycle assessment 2005 (ICLCA 2005). Int J Life Cycle Assess 10(2):163
Sonnemann G, Valdvia S (2007) First phase 2 activities of the UNEP/SETAC Life Cycle Initiative (2007–2010). Second edition of the LCA award for non-OECD countries. Jim Fava, Ana Quiros, Gregory Norris, Sonia Valdivia, Guido Sonnemann and Bas de Leeuw. Int J Life Cycle Assess 12(7):544–545
Sonnemann G, Solgaard A, Saur K, Jensen AA (2001) Life cycle management—UNEP-workshop. Sharing experiences on LCM. Int J Life Cycle Assess 6(6):325–333
Sonnemann G, Vigon B, Broadbent C, Curran MA, Finkbeiner M, Frischknecht R, Inaba A, Schanssema A, Stevenson M, Lie Ugaya CM, Wang H, Wolf M-A, Valdivia S (2011) Process on 'global guidance for LCA databases'. Int J Life Cycle Assess 16(1):95–97
Swarr TE (2006) Life cycle management and life cycle thinking—putting a price on sustainability (editorial). Int J Life Cycle Assess 11(4):217–218
Swarr TE, Hunkeler D, Klöpffer W, Pesonen HL, Ciroth A, Brent AC, Pagan R (2011) Environmental life-cycle costing—a code of practice (editorial). Int J Life Cycle Assess 16:389–391
Tak H (1999) Activities of the Korean society for life cycle assessment. Int J Life Cycle Assess 4(5):247–248
Tak H (2000) Korean society for life cycle assessment. Int J Life Cycle Assess 5(1):3
Tak H (2003) Advances of LCA activities in Korea. Int J Life Cycle Assess 8(2):63
Udo deHHA, Wrisberg N (1997a) LCA strategies—LCANET. European network for strategic life-cycle assessment research and development. Int J Life Cycle Assess 2(3):124
Udo deHHA, Wrisberg N (eds) (1997b) Life cycle assessment—state of the art and research priorities. Results of LCANET, a concerted action in the environment and climate programme (DGXII). LCA Documents vol 1. Eco-Informa Press, Bayreuth, Germany and ecomed publishers, Landsberg, Germany. ISBN 3-928379-53-4
Valdivia S, Ugaya CML, Hildenbrand J, Traverso M, Mazijn B, Sonnemann G (2012) A UNEP/SETAC approach towards a life cycle sustainability assessment—our contribution to Rio + 20. Int J Life Cycle Assess. doi 10.1007/s11367-012-0529-1
Verger G (1997) First LCA workshop in Spain 'LCA 2000'—preliminary results. Int J Life Cycle Assess 2(4):250

Verghese KL, Horne R, Carre A (2010) PIQET: the design and development of an online 'streamlined' LCA tool for sustainable packaging design decision support. Int J Life Cycle Assess 15:608–620

Wrisberg N (1998) LCA tools for decision support—Chainet. European network on chain analysis for environmental decision support. Int J Life Cycle Assess 3(2):79

Wrisberg N, Udo de Haes HA, Clift R, Frischknecht R, Grisel L, Hofstetter E, Jensen AA, Lindfors LG, Schmidt-Bleek E, Stiller H (1997) LCANET—European network for strategic life-cycle assessment research and development. A strategic research programme for life cycle assessment. Final document for the concerted action LCANET. Int J Life Cycle Assess 2(2):71–72

Ximenes FA, Grant T (2013) Quantifying the greenhouse benefits of the use of wood products in two popular house designs in Sydney, Australia. Int J Life Cycle Assess 18:891–908

Yano M, Aoki R, Nakahara Y, Itsubo N, Ohta T (2000) Current activities of the national LCA project in Japan. Int J Life Cycle Assess 5(5):261–264

Zamagni A (2012) Life cycle sustainability assessment (editorial). Int J Life Cycle Assess 17(4):373–376

Zamagni A, Amerighi O, Buttol P (2011) Strengths or bias in social LCA? (editorial). Int J Life Cycle Assess 16:596–598

Zamagni A, Pesonen H-L, Swarr T (2013) From LCA to Life Cycle Sustainability Assessment: concept, practice and future directions. Int J Life Cycle Assess 18(9):1637–1641

Chapter 6
Strengths and Limitations of Life Cycle Assessment

Mary Ann Curran

Abstract This chapter discusses strengths and limitations of Life Cycle Assessment (LCA) not by linear analysis but by elucidating limitations embedded in strengths. It elaborates perceived and real limitations in LCA methodology grouped by research need, inherent characteristic or modeling choice. So, LCA practice continues to suffer from variations in practice that can result in different LCA results. Some limitations, such as modeling missing impact indicators and making life cycle inventory more readily-available, will be addressed through continued research and development of the tool. Other modeling choice-related limitations, such as matching goal to approach setting a proper functional unit or appropriately scoping the assessment, need to be addressed through continued education and training to assist users in the proper application of the tool. Still other limitations in LCA practice would benefit by the development of harmonized guidance and global agreement by LCA practitioners and modelers.

However, despite these variations, LCA offers a strong environmental tool in the way toward sustainability.

Keywords Attributional modeling · Consequential modeling · Data uncertainty · Decision making · Functional unit · Goal and scope definition · ISO series of standards 14000 · LCA · Life cycle assessment · Life cycle impact assessment · Life cycle inventories · Life cycle sustainability assessment · Life cycle thinking · Midpoint impact categories · Modeling · Normalisation · Risk assessment · Scale · System boundaries · System expansion

1 Introduction

The last few decades have seen a marked rise in the application of life cycle assessments in virtually all countries around the world. This growing interest can be attributed to the powerful support the tool provides to decision makers. To date,

M. A. Curran (✉)
BAMAC Ltd., 5137 Oak Brook Drive,
Cincinnati, Ohio 45244, USA
e-mail: macurran@cinci.rr.com

W. Klöpffer (ed.), *Background and Future Prospects in Life Cycle Assessment,*
LCA Compendium – The Complete World of Life Cycle Assessment,
DOI 10.1007/978-94-017-8697-3_6, © Springer Science+Business Media Dordrecht 2014

Life Cycle Assessment (LCA) is a method defined by the international standards ISO 14040 and 14044 to analyse environmental aspects and impacts of product systems. In the introduction to the International Standard ISO 14040, serving as a framework, LCA is defined as follows:

> LCA studies the environmental aspects and potential impacts throughout a product´s life (i.e. cradle-to-grave) from raw material acquisition through production, use and disposal. The general categories of environmental impacts needing consideration include resource use, human health, and ecological consequences.

A similar definition of LCA was adopted as early as 1993 by the Society of Environmental Toxicology and Chemistry (SETAC) in the 'Code of Practice' document (SETAC 1993). Similar definitions can be found elsewhere. A consequence of those deliberate limitations to the analysis and interpretation of **environmental impacts** was the creation of a method that is restricted to only quantifying the **ecological** aspect of sustainability. The exclusion of economical and social factors was a deliberate choice intended to avoid method overload, while being well aware that any decision in the development of sustainable products, etc., cannot and must not neglect these factors (Klöpffer and Grahl 2014).

Among the many strengths of LCA are the following:

- *LCA is a comprehensive assessment*
  LCA is a cradle-to-grave analytical tool that captures the overall environmental impacts of all the life cycle stages associated with a product, process or human activity from raw material acquisition, through production and use phases, to waste management. This comprehensive view makes LCA a unique approach in the suite of environmental management tools available to decision makers. Without **life cycle thinking**[1], we risk focusing on the environmental issues that demand our immediate attention, and ignoring or devaluing issues that may occur either in another place or in another form (impact). Such focused assessments can lead to decisions that are based on incomplete information.
- *LCA highlights potential environmental tradeoffs*
  The broad scope involved in conducting LCA makes users more aware of the complexities of integrated industrial systems and ecosystems, and the appropriate corresponding remedy for a given situation. LCA encompasses all the interacting activities, media, and impacts and the identification of potential tradeoffs from one phase of the life-cycle to another, from one region to another, or from one environmental problem to another that may occur as a result of a decision (that is, resulting from a change to a system or from choosing between systems).
- *LCA provides structure to an investigation*
  The ISO series of standards developed in the 1990s provides us with a definition of LCA along with a general framework for conducting an assessment in four inter-related phases (goal and scope, inventory analysis, impact assessment,

[1] Life cycle thinking is a fundamental prerequisite towards understanding impact mechanisms along value chains in complex product or production systems. It is the indispensable approach to support sustainable development (De Schrynmakers 2009).

interpretation) (ISO 2006). LCA has developed into an important tool to capture information for analysis, discussion, actions and regulation in a variety of areas (Ngo 2012). LCA also assists decision makers in recognising when they intentionally or unintentionally place high value on some environmental aspects and little or no value on others.

- *LCA can challenge conventional wisdom*
The most important aspect of LCA is that it helps people incorporate whole-system thinking in terms of impact assessment. In getting away from the disconnected, stove-piped way of thinking that has led us to where we are today, LCA can bring to light data and information that makes us question what is commonly held as environmentally preferable (Ngo 2012). Bio-based materials and products, for example, have long been given preferred status. Only more recently with the reporting of LCA studies have degraded quality of water and soil resulting from biofeedstock production been brought into the discussion (von Blottnitz and Curran 2006).
- *LCA advances the knowledge base*
Taking into account the full and complete analysis of a system's environmental impacts is likely a more complicated (i.e. costly) endeavor than many organizations are willing to undertake. It is anticipated that the continued conduct of LCAs will make organisations and consumers more aware of the interconnections of operations, while providing producers, consumers and regulators with the necessary baseline information and data to move forward (Ngo 2012). The challenge now is to find an affordable, efficient way to share this growing database of knowledge with users across the globe.
- *LCA fosters communication and discourse*
The LCA methodology, originally developed to provide environmental information for distinguishing between products or between services, has evolved as a basis to communicate the overall environmental performance of products and processes to stakeholders. For example, developing environmental product declarations (EPDs) based on LCA is an effective way to communicate credible information about the environmental performance of products (Del Borghi 2012).

2 Strengths and Limitations—Perceived and Real—in Life Cycle Assessment

As with all complex assessment tools, the LCA methodology has its limitations as well as strengths. Although the ISO standard gives a consensus definition for LCA and provides a general framework for conducting an assessment, it leaves much to interpretation by the person conducting the assessment[2]. As a result, LCA studies

[2] ISO 14040 did not intend from the beginning to standardize LCA methods: "there is no single method for conducting LCA" (Heijungs and Guinée 2012).

have been criticized for producing different results for seemingly the same product. The vagueness of the ISO standard along with a growing desire to follow a 'life cycle approach' with no clear definition of what that means, has led to confusion regarding what LCA can and cannot do, and how it fits within a strategic level approach to sustainability.

Furthermore, an aspect that is simply a characteristic of LCA methodology may be perceived as a limitation if it does not fulfil the user's immediate need. For example, the present-day LCA framework does not take social welfare into consideration. Someone who is interested in understanding the social aspects of a product is recommended to apply some other tool or approach to gather information pertinent to the social (and economic) dimensions[3]. This is sometimes perceived as a missing element, or a limitation, in LCA. But it may also be viewed as an unrealistic expectation of what LCA is intended to do.

Some limitations are temporary in that the methodology could be clarified through further research and development to improve understanding of the issue and develop clear guidance. Other limitations are inherent in the design of LCA methodology and how it was intended to be conducted. Other limitations occur during application when the modeler has alternative approaches from which to choose, leading to widely varying results from case to case. In these instances, there is no 'right' way and how to approach these modeling choices is often hotly debated. LCA practice would benefit by the development of harmonized guidance and global agreement by LCA practitioners and modelers (UNEP 2011; UNEP/SETAC 2011).

Table 6.1 lists examples of LCA limitations by the three types: (1) can be improved through research; (2) inherent in the methodology; and (3) alternate modeling choices. The following sections describe these limitations in more detail.

2.1 *Matching the Goal of the Assessment to the Approach*

Not long after the 1990 SETAC workshop[4], which laid the foundation for current LCA practice, it was realized that a very important aspect had been overlooked, i.e. setting the goal for the study at the outset of the effort. Subsequent versions of the phases of an LCA in ISO included an initial 'goal and scope definition' phase (Fig. 6.1) (see Chap. 2[5] of this volume).

A clearly stated goal will make defining the study scope and data collection a little easier. For example, a study with a goal to examine bio-ethanol as an automo-

[3] Efforts to develop a Life Cycle Sustainability Assessment approach rose from the perceived need to broaden the scope of LCA from mainly environmental impacts to covering all three dimensions of sustainability (people, planet and prosperity) (CALCAS 2009). However, this broadening is at variance with ISO's explicit restriction to environmental issues (Heijungs and Guinée 2012).

[4] A Technical Framework for Life-Cycle Assessment. August 18–23, 1990, Smugglers Notch, Vermont.

[5] The role of the Society of Environmental Toxicology and Chemistry in life cycle assessment development and application by James Fava et al.

Table 6.1 Examples of limitations in LCA methodology grouped by research need, inherent characteristic or modeling choice

Research and Development to Improve LCA
Matching the goal of the assessment to the approach
Gathering the inventory data can be very resource and time intensive
Missing impact data and models for Life Cycle Impact Assessment
Dealing with life cycle inventory and impact data uncertainty
Inherent Characteristics in LCA Methodology
Distinguishing between Life Cycle Impact Assessment and Risk Assessment
LCA Does not always (usually) declare a 'winner'
LCA results should be supplemented by other tools in decision making
Choices Available to the Modeler
Allocating environmental burdens across co-products
Assigning credit for avoided burden
Expanding the boundaries (Consequential LCA)

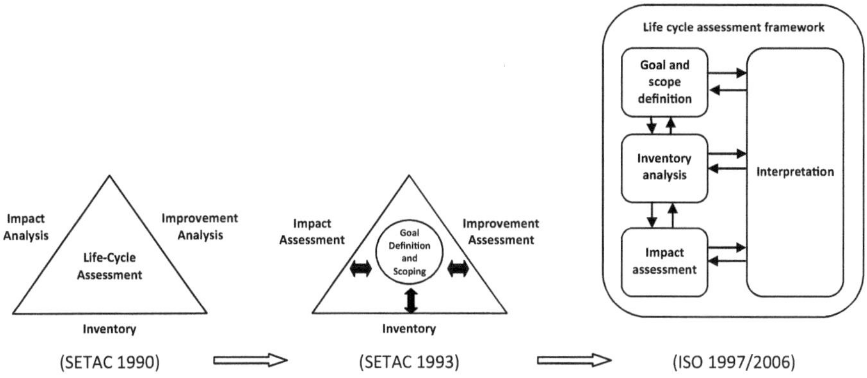

Fig. 6.1 Evolution of the LCA framework

tive fuel would lead to scoping the system around the manufacture and use of the fuel (excluding the manufacture, maintenance and end of life issues of the car itself). However, these results would then not be applicable in a comparison of, say, a car with an internal combustion engine to an all-electric vehicle, since components of the vehicle may differ (especially weight and fuel efficiency).

Although goal definition is recognized as an integral step in LCA methodology, clear guidance for matching the goal with the subsequent phases of scoping, inventory analysis and impact assessment is still lacking.

Connected with goal setting is the selection of a 'functional unit', a unique feature of LCA which sets it apart from other environmental assessment approaches. The functional unit is defined by the service provided by the system being studied. It is further shaped by the goal of the study in that it forms the basis for the study to answer the question or address the concern at hand.

Table 6.2 Functional unit versus reference flow (ISO 2012)

A functional unit is a quantified description of the performance of the product systems	Example: Lighting 10 m^2 with 3000 lx for 50,000 h with daylight spectrum at 5600 K
A reference flow is a quantified amount of manufactured product necessary for a specific product system to deliver the performance described by the functional unit	Example: 15 daylight bulbs of 10,000 lm with a lifetime of 10,000 h

At times, published LCAs report the reference flow as the functional unit and use it as the starting point for building a model of the product system; however, these two terms should not be confused. The functional unit reflects the performance or the service being fulfilled by the product system. The reference flow, then, translates the functional unit into specific product flows from the processes within the industrial system, setting the basis for calculating the inventory data (Table 6.2).

The importance of setting the appropriate scale to the functional unit was discussed early on in LCA development (e.g. Guinée et al. 2002). Often, the functional unit is set at a rather small amount; thus, the LCIA has to operate on mass loads representing a small share (often nearly infinitesimal) of the full emission output from the processes (Finnveden et al. 2009). For example, a biofuel LCA may have the functional unit of the amount of fuel to operate a single car over one year. This would require a reference flow of a small amount of biofeedstock input. The resulting impacts from the acquisition of the biofeedstock, compared to a national production level, would most likely appear insignificant, even though the potential impacts from the agricultural sector, e.g., eutrophication, land use change, soil quality, etc., may be an important consideration (Notarnicola et al. 2012). Setting the functional unit at a larger scale, such yearly production, may simplify the normalization step by giving realistic numbers for a country or an economic unit.

2.2 *Gathering the Inventory Data can be Very Resource and Time Intensive*

Although LCA databases and software have become more widely available in recent years, the lack of readily available inventory data continues to be a major hurdle for LCA practice. Inventory data can be created by collecting primary data directly from the sources, such as material and product manufacturers. More often data are collected from secondary sources such as reports, publications and databases. Data are held either privately, such as in LCA practitioners' software, or in the public domain, such as government sources. Commercial tools are usually fairly simple to use, although some training may be needed before the user is adept at using them. There is usually a subscription or purchase fee associated with these products.

While the use of readily-available software tool makes it easier to conduct an LCA, it is not always completely clear how the data were modeled in order to create the data found within them. The numerous, underlying assumptions, such as exclusions which were applied during data collection, are not typically revealed in most pre-packaged data programs. Ultimately, the user must rely on the reputation of the vendor for assurance on the quality of the data and the methods used to collect them.

Another option for creating life cycle inventories is the use of publicly-available databases. These databases are often government-sponsored, such as the US EPA's Toxic Release Inventory (TRI) and Australia's National Pollutant Inventory (NPI). They are easily accessible and available at no cost. But these sources do not lend themselves easily to use in most life cycle studies because the data are reported for individual sites or facilities and not as industry averages for a country or a region. Often assumptions have to be made about the data in order to aggregate them to represent an industry sector. Also, data are not allocated by production; therefore, additional information is needed in order to determine releases per product. To achieve this, the most effective way to simplify the LCA process is to increase the collection, publication, and standardisation of LCI data. For example, the Europeans have been successful in creating publicly-available databases through efforts such as the ecoinvent database and more recently the European Commission's Platform on Life Cycle Assessment. The US has seen limited success in creating a national inventory database (US LCI Database 2012)[6]. As mentioned earlier, it is anticipated that the continued conduct of LCAs will lead to increased generation of baseline information and data. Participation by producers, suppliers, LCA practitioners and commissioners of LCAs, in the active sharing of raw data that are collected and transformed into useful LCI data will go a long way in expanding available foreground data into the supply chain. An affordable, efficient way to share this growing database needs to be established and fully developed for public accessibility.

2.3 *Missing Impact Data and Models for LCIA*

The life cycle impact assessment (LCIA) phase is intended to provide additional information to help assess the inventory results. To do this, data that link emissions and extractions to impact categories indicators are needed. The global level models related to global warming and ozone depletion have strong agreement by LCA modelers. Other impact models are still in their infancy and in need of further development, such as water use, land use, and in addressing issues such as spatial and temporal differentiation (Margni and Curran 2012). While both abiotic and biotic resources are generally considered to be equally important, modeling biotic resource use has not received as much attention (Finnveden et al. 2009).

Further yet, some impact data are yet to be generated and made publicly-available. For example, impact data for human and ecosystem health exposure to nanoproducts (products that contain a nanocomponent or produced using nanotechnol-

[6] See the 'US LCI Database Project—Review Panel Report on the Development Guidelines' from January 2004 (www.nrel.gov/lci/pdfs/34275.pdf). NREL is a national laboratory of the U.S. Department of Energy, Office of Energy Efficiency and Renewable Energy operated by the Alliance for Sustainable Energy, LLC.

ogy) are still insufficient. Another example involves modeling the management of nuclear waste from nuclear power generation. In both cases, current LCIA models cannot fully model the inventory data for these systems; the modeler runs the risk of dropping important inventory data if it they are not otherwise retained and reported in the final analysis.

Currently, there is no one single impact assessment methodology being used by practitioners. Nevertheless, commonalities can be seen in LCA practice regarding the impact categories that are being selected for modeling. Table 6.3 lists midpoint impact categories that are being used by prominent researchers in their LCIA models[7].

2.4 Dealing with Data Uncertainty

Uncertainty analysis is the process of determining the variability of the data and the impact on the final results. It applies to both the inventory data and the impact assessment indicators and can be attributed to both errors and normal fluctuations in the data. While data variability can have a great impact on how the results are used in decision-making, the actual influence of uncertainty on decision-making has not been adequately studied. Furthermore, many LCAs are produced without reporting the uncertainty of the data. There is a need to understand the consequences of these decisions for proper transparency in the study.

Research efforts are needed to establish recommended practice for uncertainty analysis and to elaborate guidance for practitioners and method developers on how to estimate, communicate, interpret and manage uncertainty in both LCI and LCIA (Margni and Curran 2012).

2.5 Distinguishing between Life Cycle Impact Assessment and Risk Assessment

It is important to understand the difference between Life Cycle Impact Assessment (LCIA) methodology and traditional Risk Assessment (US EPA 2004). The general approach to risk assessment is a complex process, requiring the integration of data and information across a broad range of activities and disciplines, including source characterisation, fate and transport, modeling, exposure assessment, and dose-response assessment. On the other hand, in an LCA the product system is extended in space and time, and the emission inventory is often aggregated in a form which restricts knowledge about the geographical location of the individual emissions. The

[7] While midpoint modeling is most common in LCA practice, some methods model past the midpoint to the endpoint level (e.g., from an ozone depletion indicator to increased incidents of skin cancer). These damage models can be reported in units of Disability Adjusted Life Years (DALYs), an aggregation of environmental impacts, monetary value, or other aggregated damage units.

Table 6.3 Midpoint Impact Categories included in ReCiPe, IMPACT World+, TRACI and LIME (Margni and Curran 2012)

ReCiPe		IMPACT World+	TRACI	LIME
http://www.lcia-recipe.net/		http://www.impactworldplus.org	http://www.epa.gov/nrmrl/std/traci/traci.html	http://www.jemai.or.jp/lcaforum/index.cfm
INPUTS	Mineral Resource Depletion Fossil Fuel Depletion	Resource Use	Resource Depletion Fossil Fuel Use	Resource Consumption
	Agricultural Land Occupation Urban Land Occupation Natural Land Transformation	Land Use	Habitat/T&E Species	Land Use
	Water Depletion	Water Use	Water Use	
OUTPUTS	Climate Change	Global Warming	Global Warming	Global Warming
	Ozone Depletion	Ozone Layer Depletion	Ozone Depletion	Ozone Layer Depletion
	Particulate Matter Formation			Urban Air Pollution
	Human Toxicity	Human Toxicity	Human Health – Cancer – Non Cancer – Criteria Pollutants	Human Toxicity
	Terrestrial Ecotoxicity Freshwater Ecotoxicity Marine Ecotoxicity	Ecotoxicity	Ecotoxicity	Ecotoxicity
	Photochemical Oxidant Formation	Photochemical Oxidation	Smog Formation	Photochemical Oxidant
	Terrestrial Acidification	Acidification	Acidification	Acidification
	Freshwater Eutrophication Marine Eutrophication	Eutrophication	Eutrophication	Eutrophication
	Ionising Radiation			Waste

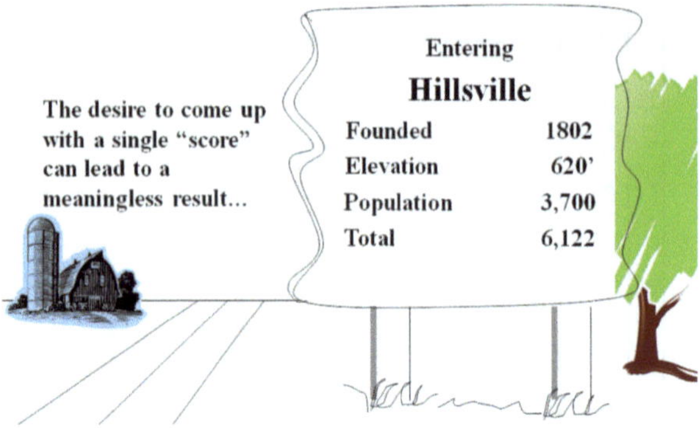

Fig. 6.2 Collapsing different impact category indicators into a single score is a subjective process involving weighting and normalization

LCI results are also typically unaccompanied by information about the temporal course of the emission (some environmental impacts may occur in the future) or the resulting concentrations in the receiving environment (Finnveden et al. 2009). With the inherent uncertainty in modeling environmental impacts, an impact indicator is the outcome of a simplified model of a very complex reality, giving only an approximation of the quality status of the affected entity. If not sufficient for absolute predictions of risk, LCIA models and LCA results are suitable for assessing relative comparisons.

2.6 LCA Does not Always (usually) Declare a 'Winner'

Converting impact results to a single score is a subjective process requiring value judgments[8], which cannot be based solely on natural science. All assumptions or decisions made throughout the study must be reported. If not, the final results may be taken out of context or misinterpreted (Fig. 6.2).

The interpretation phase of LCA entails the evaluation of the results of the inventory analysis along with the results of the impact assessment to aid in the decision making process, whether it is to select the preferred product, improve a process or service, etc. with a clear understanding of the uncertainty and the assumptions used to generate the results. Very seldom will the results of an LCA identify a clear 'winner' between alternatives. In some cases, it may not be possible to state that one alternative is better than the others because of the uncertainty in the final results. This does not imply that efforts have been wasted or that LCA is not a viable tool for decision makers. The LCA process will still improve understanding of the environmental and health impacts associated with each alternative, where they occur

[8] Value judgments include the application of weighting (assignment and calculation of different impact categories and resources reflecting their relative importance) and normalisation (calculation of the magnitude of the category indicator results). In the ISO standard, normalisation is allowed for comparative assertions intended to be made available to the public, but not weighting due to its inherently subjective nature (ISO 14040+44).

(locally, regionally, or globally), and the relative magnitude of each type of impact in comparison to each of the proposed alternatives included in the study. This information more fully reveals the pros and cons of each alternative.

LCA can be used to establish a baseline of a product's environmental profile. But it is best used as a relative tool intended for comparison, and not absolute evaluation, thereby helping decision makers compare all major environmental impacts when choosing between alternative courses of action.

2.7 LCA Results should be Supplemented by Other Tools in Decision Making

While an LCA study produces very useful information, the results should be used as one component in a comprehensive decision-making process. It may be necessary to supplement the LCA with other tools or methods to provide a basis for decision-making. These tools include risk assessment, site-specific environmental assessment, cost assessment and others. As a part of the scoping process, it is useful to identify where and how these other tools will be used to augment the findings of the LCA. Further development is needed to create an integrated framework to reduce complexity while clarifying the simplification choices which have been made in the integrative analysis (CALCAS 2008).

In addition, the nature of LCA as an iterative process is often overlooked. Interpretation of the findings is about comparing the data and results with previous findings, and putting them in the proper context of decision-making and limitations. The iterative nature of the ISO framework (see Chap. 3[9] of this volume) shows up in this context. If the uncertainties are too high, we may go back to collect better data. If the sensitivity analysis shows that some decisions are crucial, we may go back and do a more refined analysis. It is especially important to determine that if the results of the impact assessment or the underlying inventory data are incomplete or unacceptable for drawing conclusions and making recommendations, then the previous steps must be repeated until the results can support the original goal of the study.

The decision tree shown in Fig. 6.3 depicts an iterative approach to collecting information in support of the decision making process for nanoproduct development (US EPA 2011). This approach, which follows the 'three pillar' interpretation of sustainability, can be applied to any product.

2.8 Allocating Environmental Burdens Across Co-products

When a process makes multiple products, the question of how to assign material use and environmental releases to each co-product becomes relevant. The ISO standard

[9] The international standards as the constitution of life cycle assessment: the ISO 14040 series and its offspring by Matthias Finkbeiner.

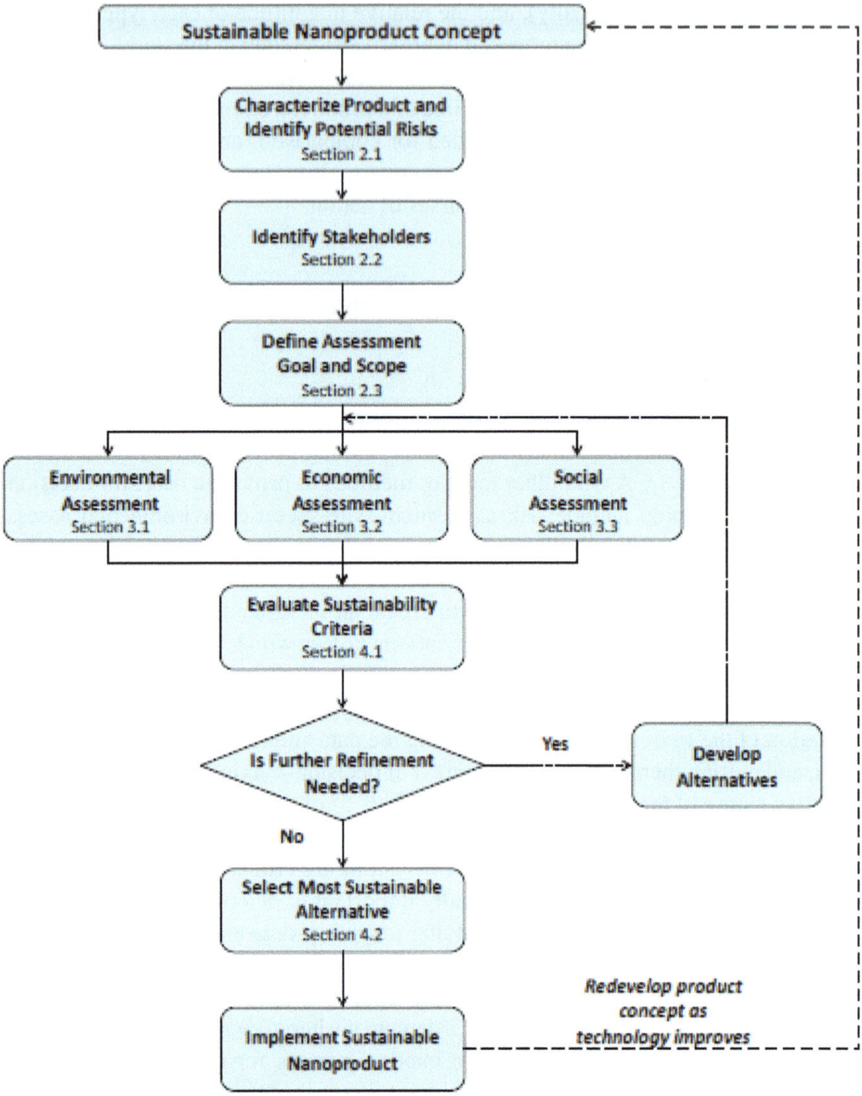

Fig. 6.3 US EPA's framework for sustainable nanotechnology. (US EPA 2011)

provides some guidance in the form of a hierarchy (Box 1), which calls for practitioners to avoid allocation if possible, by either (1) Modeling the sub-processes involved in production (i.e. collect more detailed data), or (2) Expanding the system boundaries to include additional processes that relate to the co-product(s). But much is left to interpretation in practice.

Box 1 Co-product allocation hierarchy (ISO 2006)

ISO 14041 6.5.3

On the basis of the principles mentioned above, the following step-wise procedure shall be applied.

Step 1: Wherever possible, allocation should be avoided by:

1. Dividing the unit processes to be allocated into two or more subprocesses and collecting the input and output data related to these subprocesses.
2. Expanding the product system to include the additional functions related to the co-products, taking into account the requirements of (function, functional unit and reference flow).

Step 2: Where allocation cannot be avoided, the inputs and outputs of the system should be partitioned between its different products or functions in a way which reflect the underlying physical relationships between them, i.e. they shall reflect the way in which the inputs and outputs are changed by quantitative changes in the products or functions delivered by the system. The resulting allocation will not necessarily be in proportion to any simple measurement such as mass or molar flows of co-products.

Step 3: Where physical relationship alone cannot be established, or used as the basis for allocation, the inputs should be allocated between the products and functions in a way which reflects other relationships between them. For example, input and output data might be allocated between co-products in proportion to the economic value of the products.

There is general agreement that avoiding allocation through sub-process modeling and system expansion (Step 1 of the ISO hierarchy) is an appealing way to handle this seemingly intractable problem. However, both approaches cause the model to get larger and more complicated, requiring the collection of more data in order to complete the analysis. Collecting more data means more time and effort which brings the practicality of the approach into question. Also, larger systems run the risk of being less transparent in that there is more information on how the data were arrived at than can be easily communicated. So, although the answers that would be obtained through sub-process modeling would be more relevant to sustainability and more useful in helping decision-makers make better decision, allocation may not always be avoidable, especially if the data for the sub-processes or for the expanded system cannot be easily acquired.

2.9 Assigning Credit for Avoided Burden

In a system expansion approach, the boundaries are expanded to include the alternative production of exported functions. To do this, a necessary requirement of system expansion is the existence of an alternative way to produce a by-product. While this

Table 6.4 Energy ratio to produce corn ethanol calculated with co-product credit, 1996 (USDA 2002)

	Ethanol	Co-Products	Energy Use without Co-Product Credit	Energy Use with Co-Product Credit	NEV with Co-Products	Energy Ratio
	Percent	Percent	Btu/gal	Btu/gal[7]	Btu/gal[7]	Btu/gal[7]
Output weight basis:						
Wet mill	48	52	79,503	38,987	44,974	2.15
Dry mill	49	51	74,447	37,289	46,672	2.25
Weighted average	48	52	77,228	37,895	46,066	2.22

1000 Btu/US gallon = 0.279 megajoules per liter (MJ/l)
NEV Net Energy Value

concept seems reasonable on the surface, it can be controversial. It is often used to 'credit' the system with avoided burdens that are offset by the alternative process.

For example, corn mills produce both ethanol and corn oil; the corn ethanol system can be credited with the amount of energy it would have taken to make a competing product, such as soybean oil (Table 6.4). Not only does system expansion require more data to be collected, it also presents a problem with conveying the results of the study depending upon how the process in question was modeled. It is easy to see how the application of system expansion can have a significant impact on the study results.

Recycling, specifically open-loop recycling, is viewed as a special condition of allocation and is given special attention in the literature. The concern is to capture the downstream costs and benefits that post-consumer recycling may incur. Economic allocation seems to be the preferred approach and is perceived to be the best avenue to capture the downstream recycling activities. A number of allocation methods for open loop recycling are based on arguments about fairness, or accountability, so that environmental burden is appropriately assigned to the offending activity. However, it is difficult to determine which procedure is most 'fair' since this is a subjective term and depends on the perspective of the person conducting the study. The ISO 14040+44 standards stipulate the conduct of sensitivity analysis if "subjective" allocations are applied in order to show the effect the choice has on the results.

2.10 Expanding the Boundaries (consequential LCA)

By 2005, LCA practitioners began to take notice of expanded study boundaries that encompass the likely consequences of change resulting from a decision. This expanded approach to LCA became known as consequential LCA (Curran et al. 2005) to distinguish it from the more system-confined approach of attributional LCA. The change in the balance between supply and demand for a good or service can have a far-reaching impact. For example, Searchinger et al. (2008) found an attributional analysis of US corn-based ethanol resulted in a 20% *decrease* in greenhouse gas emissions compared to conventional gasoline. However, in a consequential analysis to account for policy-driven increases in output, they predicted a 47% *increase* in emissions compared to gasoline, due to land use changes induced by higher prices of corn, soybeans and other grains from anticipated additional demand for corn starch for ethanol production.

It is possible that the inventory results of a consequential LCA will be negative, if the change in the level of production causes a reduction in emissions greater than the emissions from the production of the product. This does not mean that the absolute emissions from the production of the product are negative, but that the production of the product will cause a reduction in emissions elsewhere in the system (Ekvall et al. 2005).

A consequential LCA is conceptually complex because it includes additional, economic concepts such as marginal production costs, elasticity of supply and demand, etc. A report prepared for the project 'Co-ordination action for innovation in life-cycle analysis for sustainability' (CALCAS 2009) outlines a four-step procedure to identify which unit processes to link:

- Identifying the scale and time horizon of the potential change studied;
- Identifying the limits of a market;
- Identifying trends in the volume of a market; and
- Identifying changes in supply and demand.

Therefore, consequential LCA depends on descriptions of economic relationships embedded in models. It generally attempts to reflect complex economic relationships by extrapolating historical trends in prices, consumption and outputs. This adds to the risk that inadequate assumptions or other errors significantly affect the final LCA results. To reduce this risk, it is important to ensure that the various results regarding different consequences can be explained using credible arguments. The main limitation for applying consequential LCA is the lack of the data in current LCA databases needed to support this type of modeling (CALCAS 2009).

There is no right or wrong choice between the attributional and consequential approaches, and the ISO standard does not offer specific guidance on how the goal of the study affects the scoping of the system boundary. While consequential modeling is relevant in most application areas of LCA, there are applications where the typical decisions studied by LCA are not of such significant size[10] and attributional modeling could be considered (CALCAS 2008). The distinction between attributional and consequential LCA is one example of how choices in the Goal and Scope Definition of an LCA influence methodological and data choices for the LCI and LCIA phases (Finnveden et al. 2009).

[10] A decision is considered small or marginal when it does not affect the determining parameters of the overall market situation, that is, the direction of the trend in market volume and the constraints on and production costs of the involved products and technologies (CALCAS 2009).

Table 6.5 Life-cycle based approaches with a single issue focus

Title	Impact Metric
Life Cycle Greenhouse Gas (GHG) Analysis: 'Direct' as well as 'indirect' GHG emissions across the product lifecycle	Global Warming
Carbon Footprint: 'Direct' emissions of carbon dioxide (CO_2) from burning fossil fuels including domestic energy consumption and transportation as well as of 'indirect' CO_2 emissions from the product lifecycle	Global Warming
Water Footprint: Freshwater used by individuals or organisations to make goods or provide services	Water Depletion
Ecological Footprint: The amount of cropland, grazing land, forest area, and fishing grounds needed to satisfy a population's need for food, clothing, shelter, products and services, plus the amount of land required to absorb wastes	Land and Resource Use
Net Energy Balance: The overall gain or loss of energy, measured typically in Joules	Energy Production and Use
Chemical (Risk) Life Cycle: Multi-media environmental fate and transport, exposure, and effects on ecological receptors and human health across the life cycle of a chemical	Human and Ecological Health

3 Life Cycle Thinking

The preceding sections address issues related to the ISO-defined LCA methodology. In recent years the growing popularity of LCA and the life cycle concept have led to simplified approaches that focus on a single impact, thereby reducing the effort needed for data collection, impact assessment, and reporting. Table 6.5 lists several life cycle-based approaches that are commonly used to analyse and report select impact metrics.

It is clear that there is much variability in what life cycle-based tools measure and report (Curran 2013). In contrasting these approaches against LCA, it is also clear that focusing on specific issues of concern and not considering the whole suite of potential environmental concerns, risks overlooking potential burden shifting that may occur as a result of a decision. The conduct of an assessment that models only one or two pre-selected impact categories does not meet the definition of LCA, according to the ISO standards 14040+44.

4 Conclusion

Increasingly, decision makers are turning to LCA as a proven methodology to assess potential environmental impacts of products, goods and services. The ISO 14000 standard series provides a broadly accepted set of principles and the present-day

LCA framework. While LCA has come a long way in the development of methodology and continues to evolve with additional knowledge, LCA practice continues to suffer from variations in practice that can result in different LCA results. Some limitations, such as modeling missing impact indicators and making life cycle inventory more readily-available, will be addressed through continued research and development of the tool. Other modeling choice-related limitations, such as matching goal to approach setting a proper functional unit or appropriately scoping the assessment, need to be addressed through continued education and training to assist users in the proper application of the tool. Still other limitations in LCA practice would benefit by the development of harmonized guidance and global agreement by LCA practitioners and modelers.

Despite the variations outlined previously, LCA offers a strong environmental tool in our journey toward sustainability. Meeting the challenge of shifting the paradigm to one where LCA is the foundation of decision-making in regulation and commerce depends on public and private policy makers changing their belief systems and behaviors so their choices serve both current and future generations (Ngo 2012).

References

von Blottnitz H, Curran MA (2006) A review of assessments conducted on bio-ethanol as a transportation fuel from a net energy, greenhouse gas, and environmental life cycle perspective. J Clean Prod 15(7):607–619

Del Borghi A (2012) Editorial: LCA and communication-environmental product declaration. Int J Life Cycle Assess 18:293–295

CALCAS (2008) Critical review of the current research needs and limitations related to ISO LCA practice guidelines for application of deepened and broadened LCA. Co-ordination action for innovation in life-cycle analysis for sustainability, Project No. 037075, Deliverable 20 of workpackage 7 prepared by Zamagni A, Buttol P, Porta PL, Buonamici R, Masoni P, Guinée J, Heijungs R, Ekvall T, Bersani R, Bień kowska A, Pretato U

CALCAS (2009) Guidelines for application of deepened and broadened LCA. Co-ordination action for innovation in life-cycle analysis for sustainability, Project No. 037075, prepared by Pedersen B, Ekvall T, Heijungs R

Curran MA, Mann M, Norris G (2005) The international workshop on electricity data for life cycle inventories. J Clean Prod 13(8):853–862

Curran MA (2013) Assessing environmental impacts of biofuels using life cycle-based approaches. Manag Environ Qual 24(1):34–52

De Schrynmakers P (2009) Life cycle thinking in the aluminium industry. Int J Life Cycle Assess 14(Suppl 1):S2–S5

Ekvall T, Tillman A-M, Molander S (2005) Normative ethics and methodology for life cycle assessment. J Clean Prod 13(13–14):1225–12334

Finnveden G, Hauschild MZ, Ekvall T, Guinée J, Heijungs R, Hellweg S, Koehler A, Pennington D, Suh S (2009) Recent developments in life cycle assessment. J Environ Manag 91:1–21

Guinée JB, Gorrée M, Heijungs R, Huppes G, Kleijn R, Koning A de, Oers L van, Wegener Sleeswijk A, Suh S, Udo de Haes HA, Bruijn H de, Duin R van, Huijbregts MAJ (2002) Handbook on life cycle assessment. Operational guide to the ISO standards. I: LCA in perspective. IIa: Guide. IIb: Operational annex. III: Scientific background. Kluwer Academic Publishers, Dordrecht, 692 pp, ISBN 1-4020-0228-9

Heijungs R, Guinée J (2012) An overview of the life cycle assessment method-past, present, and future. In: Curran MA (ed) Life cycle assessment handbook: a guide for environmentally sustainable products, Chapter 2. Scrivener Publishing, Beverly

ISO (1997) ISO 14040:1997(E) Environmental management—Life cycle assessment—Principles and framework. International Organisation for Standardisation, Geneva

ISO (2006) ISO 14040:2006(E) Environmental management—Life cycle assessment—Principles and framework. International Organisation for Standardisation, Geneva

ISO (2012) ISO/TR 14049:2012 Environmental management—Life cycle assessment—Examples of application of ISO 14041 to goal and scope definition and inventory analysis. International Organisation for Standardisation, Geneva

Klöpffer W, Grahl B (2014) Life cycle assessment (LCA)-a guide to best practice. Wiley-VCH, Weinheim

Margni M, Curran MA (2012) Life cycle impact assessment. In: Curran MA (ed) Life cycle assessment handbook: a guide for environmentally sustainable products, Chapter 4. Scrivener Publishing, Beverly

Notarnicola B, Tassielli G, Renzulli P (2012) Modeling the GRI-food industry with life cycle assessment. In: Curran MA (ed) Life cycle assessment handbook: a guide for environmentally sustainable products, Chapter 7. Scrivener Publishing, Beverly

Ngo AK (2012) Environmental accountability: a new paradigm for world trade is emerging. In: Curran MA (ed) Life cycle assessment handbook: a guide for environmentally sustainable products, Chapter 24. Scrivener Publishing, Beverly

Searchinger T, Heimlich R et al (2008) Use of US cropland for biofuels increases greenhouse gases through emissions from land-use change. Science 319(5867):1238–1240

SETAC (1990) A technical framework for life cycle assessments. In: Fava J, Denison R et al (eds) The Society of Environmental Toxicology and Chemistry. Brussels

SETAC (1993) Guidelines for life-cycle assessment: a code of practice. In: Consoli F, Allen D et al (eds) Society of Environmental Toxicology and Chemistry. Brussels

UNEP (2011) Global guidance principles for life cycle assessment databases: a basis for greener processes and products. United Nations Environment Programme, Paris

UNEP/SETAC (2011) Valdivia S, Ugaya CML, Sonnemann G, Hildenbrand J (eds) Towards a life cycle sustainability assessment-making informed choices on products. ISBN 978-92-807-3175-0 Paris (http://lcinitiative.unep.fr)

USDA (2002) The energy balance of corn ethanol-an update. By Shapouri H, Duffield JA, Wang M. U.S. Department of Agriculture, Office of the Chief Economist, Office of Energy Policy and New Uses. Agricultural Economic Report no 813

US EPA (2004) An examination of EPA risk assessment principles and practices, EPA/100/B-04/00. Office of the Science Advisor. Washington, DC

US EPA (2011) Guidance to facilitate decisions for sustainable nanotechnology, EPA/600/R-11/107. US Environmental Protection Agency. Office of Research & Development. Cincinnati

Database USLCI (2012) National renewable energy laboratory. https://www.lcacommons.gov/nrel/search. Accessed 19 Nov 2012

Chapter 7
Challenges in Life Cycle Assessment: An Overview of Current Gaps and Research Needs

Matthias Finkbeiner, Robert Ackermann, Vanessa Bach, Markus Berger, Gerhard Brankatschk, Ya-Ju Chang, Marina Grinberg, Annekatrin Lehmann, Julia Martínez-Blanco, Nikolay Minkov, Sabrina Neugebauer, René Scheumann, Laura Schneider and Kirana Wolf

Abstract This chapter provides a comprehensive overview of current gaps of and challenges for LCA structured into inventory, impact assessment, generic and evolving aspects. A total of 34 gaps and challenges were identified. These include challenges like 'allocation', 'uncertainty' or 'biodiversity', as well as issues like 'littering', 'animal well-being' or 'positive impacts' which are not covered as often in the existing LCA literature. Each of these gaps is described by a high-level overview of the topic and its relevance to LCA, and the state of the art in terms of literature and potential solutions, if any, is presented.

The motivation for such an overview is two-fold: First, robust, sustainable and credible use of LCA should avoid the over-interpretation of LCA results without proper consideration of its gaps and limitations. Second, these gaps and challenges represent research needs for the scientific LCA community and hopefully inspire further progress in method development.

Keywords Abiotic resources · Allocation · Animal well-being · Biodiversity · Biogenic carbon · Biological invasion · Biotic resources · Soil quality · Consequential LCA · Data quality analysis · Decision making · Delayed emissions · Desertification · Ecosystems · Ecotoxicity · Effects of chemicals · Functional unit · Human health · Human toxicity · Impact assessment · Improbable events · Inventory analysis · ISO 14040 · ISO 14044 · Land use · Life cycle assessment (LCA) · Limitations · Littering · Macroeconomic scale-up · Noise · Odor · Positive impacts · Rebound effects · Renewable energy · Research needs · Resources · Robustness in LCA · Salinization · Stakeholders · Uncertainty analysis · Uncertainty · Value choices · Water use and consumption · Weighting

---

M. Finkbeiner (✉) · R. Ackermann · V. Bach · M. Berger · G. Brankatschk · Ya-Ju Chang · M. Grinberg · A. Lehmann · J. Martínez-Blanco · N. Minkov · S. Neugebauer · R. Scheumann · L. Schneider · K. Wolf
Department of Environmental Technology, Chair of Sustainable Engineering, Technische Universität Berlin, Office Z1, Strasse des 17. Juni 135, 10623, Berlin, Germany
e-mail: matthias.finkbeiner@tu-berlin.de

W. Klöpffer (ed.), *Background and Future Prospects in Life Cycle Assessment*, LCA Compendium – The Complete World of Life Cycle Assessment,
DOI 10.1007/978-94-017-8697-3_7, © Springer Science+Business Media Dordrecht 2014

1 Introduction

The good news is: life cycle assessment (LCA) is approaching mainstream. After many years of method development, case studies, international standardization, database and software development, LCA is mature and robust enough to be used for decision-making—in both private and public organizations. LCA is currently the most accepted tool to assess the environmental performance of products and this basically applies all around the globe and to all stakeholders, e.g. government, industry, non-governmental organizations (NGOs), academia.

Ten years ago, the European Commission testified in their Communication on Integrated Product Policy (IPP), that LCA is the "…best framework for assessing the potential environmental impacts of products currently available" (EU 2003). This statement stood the test of time and is nowadays common sense beyond Europe. However, being 'the best available method' does not mean that LCA is 'perfect'. While the LCA community had to promote LCA uptake for many years, it is now important not to oversell it. A balanced understanding and use of LCA is needed for ensuring sustainable success. We have to avoid going from the 'LCA-aversion' of the past straight into a kind of 'LCA-hype'.

Both, the international standards of LCA and the scientific literature are quite transparent with regard to the gaps and challenges of the method. LCA does not provide the 'full environmental truth', at least not just yet. The core standards of LCA (ISO 14040 2006) and (ISO 14044 2006) (see Chap. 3, this volume, entitled 'The international standards as constitution of LCA: the ISO 14040 series and its offspring') acknowledge clearly that any LCA study has its limitations. Therefore, the limitations of every study have to be documented in the goal and scope definition (ISO 14040, 5.2.1.2). Moreover, there is a specific chapter on limitations of life cycle impact assessment (LCIA), i.e. ISO 14040, 5.4.3 Limitations of LCIA: "The LCIA addresses only the environmental issues that are specified in the goal and scope. Therefore, LCIA is not a complete assessment of all environmental issues of the product system under study. LCIA cannot always demonstrate significant differences between impact categories and the related indicator results of alternative product systems. This may be due to

- limited development of the characterization models, sensitivity analysis and uncertainty analysis for the LCIA phase,
- limitations of the LCI [life cycle inventory] phase, such as setting the system boundary, that do not encompass all possible unit processes for a product system or do not include all inputs and outputs of every unit process, since there are cut-offs and data gaps,
- limitations of the LCI phase, such as inadequate LCI data quality which may, for instance, be caused by uncertainties or differences in allocation and aggregation procedures, and
- limitations in the collection of inventory data appropriate and representative for each impact category" (ISO 14040 2006).

In the scientific literature, an early analysis of drawbacks was performed by Udo de Haes (Udo de Haes 1993). More recent contributions with regard to gaps and research needs include Reap et al. (2008a, b), Finnveden et al. (2009) and Klöpffer and Grahl (2009). However, based on the understanding of this decade, a comprehensive overview of gaps and challenges is still missing, especially reflecting the more recent developments with regard to carbon footprinting (Finkbeiner 2009) and water footprinting (Berger and Finkbeiner 2010) as well as the less explored aspects like littering or animal well-being.

The following section details the approach and methodology chosen for this review article. Section 3 presents the resulting gaps, challenges and research needs.

2 Methodology

This contribution is based on an extensive desk and literature research by an experienced and interdisciplinary group of LCA scientists and practitioners. Each of the co-authors was responsible for a set of topics, performed the associated literature survey and prepared the necessary background material. Due to the large number of gaps and challenges identified, it was necessary to restrict the gap description and analysis to a high-level summary. More detailed and comprehensive reflections on individual gaps go beyond the scope of this contribution.

The comprehensiveness of the selection of challenges was not stretched to the limit. There are even further issues which could have been included as challenges for LCA. As examples, normalization, definition of system boundaries or the application of cut-off criteria are not discussed, because they were comprehensively covered in previous reviews.

For consistency in the gap descriptions, a common format is used to describe each gap. Four guiding questions were analyzed for every challenge. In order to support readability and easy reference, the presentation in the result section is organized in such a way that—whenever feasible—the guiding questions are discussed in a respective paragraph and according to the following order:

1. What is the topic about?
2. Why is it a gap of or challenge for LCA and in which case is it particularly relevant?
3. What is the state of the art in the scientific literature?
4. What can be done to address the gap or challenge?

Basically, all gap descriptions are self-sufficient. However, to provide a structure for their presentation and for readability purposes they were attributed to one of the following topics which correspond with individual subsections in the results part of the chapter:

- Inventory aspects (see Sect. 3.1)
- Impact assessment aspects (see Sect. 3.2)
  - Human health (see Sect. 3.2.1)
  - Ecosystem (see Sect. 3.2.2)
  - Resources (see Sect. 3.2.3)

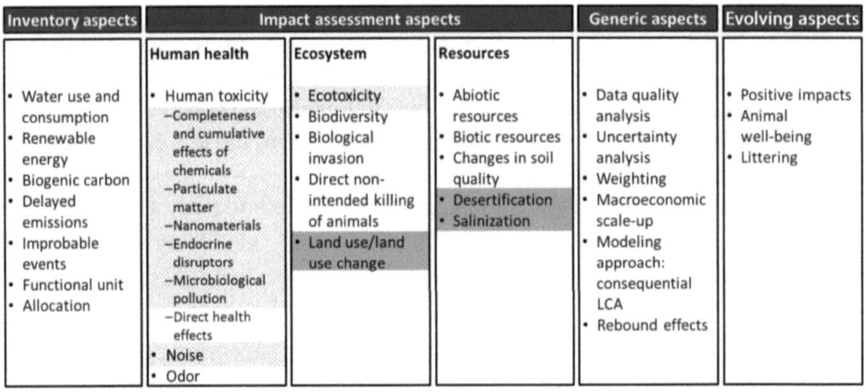

Fig. 7.1 Structured overview of challenges for LCA (shaded areas indicate crosscutting between areas of protection within impact assessment)

- Generic aspects (see Sect. 3.3)
- Evolving aspects (see Sect. 3.4)

It should be noted that the focus of this contribution is clearly environmental LCA and does not include a discussion of gaps and challenges of social LCA (SLCA) (UNEP/SETAC 2009), life cycle costing (LCC) (Hunkeler et al. 2008; Swarr et al. 2011) or comprehensive life cycle sustainability assessment (LCSA) (Klöpffer 2008; Finkbeiner et al. 2010; UNEP/SETAC 2011b).

3 Results

As a main result of this contribution 34 methodological gaps of and challenges for LCA were identified and grouped according to the structure presented in the section above. Figure 7.1 provides an overview of the topics covered.

In several cases, the attribution of gaps and challenges to one of the overall topics is not straightforward. In such ambiguous cases, the decision for a certain topic is based on our judgment as to which of them represents the most dominant aspect of the gap. As an example, there are definitely still challenges with regard to the impact assessment of 'water use and consumption' (see Sect. 3.1.1), but the currently limiting factor in application of advanced water footprint assessment is the lack of proper inventory data. Therefore, the challenge 'water use and consumption' is covered under inventory aspects.

Because of the larger number of challenges and the availability of established areas of protection, the impact assessment aspects are further differentiated into human health, ecosystem and resource topics (Fig. 7.1). Within impact assessment some of the gaps and challenges are clearly crosscutting between areas of protection. This is indicated by the shaded areas in Fig. 7.1. As an example, 'nanomaterials', 'endocrine disruptors' and 'noise' are attributed and discussed first in the

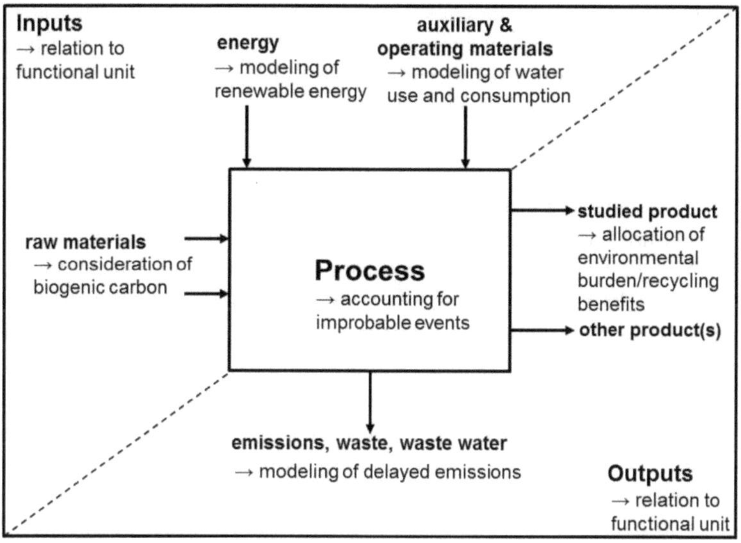

Fig. 7.2 Overview of process inputs and outputs (adapted from Klöpffer and Grahl 2009) and relation to challenges referring to 'inventory aspects' of LCA

human health section, but they are equally relevant for ecosystems. The following sections provide the individual gap descriptions according to the structure presented in Fig. 7.1.

3.1 Inventory Aspects

This section analyzes gaps and challenges referring to inventory aspects of LCA. This includes a discussion on 'water use and consumption', 'renewable energy', 'biogenic carbon' and 'delayed emissions' as well as the inclusion of 'improbable events' in LCA. Moreover, inherent challenges like 'allocation' and 'functional unit' are discussed.

In Fig. 7.2 gaps and challenges regarding the life cycle inventory are presented and the relation to inputs and outputs of modeled processes in LCA is illustrated.

3.1.1 Water Use and Consumption

While water use is the total freshwater input into a product system, water consumption denotes the fraction of water use which is not returned to the originating river basin, mainly due to evapo(transpi)ration or product integration (Bayart et al. 2010).

Some consequences of water use, e.g. eutrophication or human- and ecotoxicity, are sufficiently covered in LCIA by respective impact categories (e.g. Guinée et al. 2002). Additionally, several inventory and impact assessment methods were developed, describing various cause-effect chains on human health, ecosystems, and resources (Berger and Finkbeiner 2010). Since water scarcity is a local phenomenon, impacts of water consumption need to be assessed on a local level, too. Therefore, all impact assessment methods require regionalized inventory flows. Additionally, some methods need information concerning types of water courses, water qualities, and time of consumption. Despite great progress in method development, hardly any of these methods have been applied in practice as inventory requirements are hard to satisfy—especially if complex background systems are involved. So the greatest challenge regarding water use and consumption assessment in LCA is the lack of detailed inventory information and application of existing impact assessment methods. Considering consequences of water consumption is especially relevant for LCAs of agricultural products, like food, natural fibers, or biofuels, as the agricultural sector is responsible for 85% of global water consumption (Shiklomanov 2003). As energy production is the largest water consumer in industry (Pfister et al. 2011), water consumption should be assessed in energy intense industrial product systems, too.

In order to overcome the gap of lacking regionalized water inventories, Berger et al. (2012) developed a method to regionalize aggregated water consumption figures of LCI databases. In a top-down-approach, the total water consumed in the production of three passenger cars was allocated to manufacturing processes and material groups. Then the material-specific water consumption was assigned to countries based on import mix shares, locations of production sites and suppliers, etc. In terms of water quality, Boulay et al. (2011) developed inventory categories allowing for user-friendly consideration of water quality aspects.

In the short term, it is recommended to collect or estimate local water inventory data as detailed as possible and to apply existing impact assessment methods. In the long term, regionalized water inventory flows and impact assessment methods for water consumption have to be implemented into LCI databases and software tools.

In this context it should be noted that the challenge of lacking regionalized inventory data is also relevant for other impact categories like land use (Sect. 3.2.2.5) or biodiversity (Sect. 3.2.2.2). In these cases similar recommendations on the collection of inventory data and development of databases apply.

3.1.2 Renewable Energy

Renewable energy comes from resources which are continuously replenished, such as biomass, wind, water, earth (geothermal energy), and sun.

Several LCA studies are available for different types of renewable energy systems (Pehnt 2006; Dovì et al. 2009; Varun et al. 2009; Ardente et al. 2005; Burkhardt et al. 2012; Hsu et al. 2012; Ardente et al. 2008; Schleisner 2000; Arvesen and Hertwich 2012; Yee et al. 2009; Cherubini et al. 2009; Dodić et al. 2010). However, if renewable energy is used as an input in other product systems, it is still unclear how it can be modeled in a robust and consistent way. The two main issues are the potential double-counting of renewable energy by modeling specific uses of renewable energy that are considered at the same time in the grid mix and the question whether the production mix or consumption mix of a grid is modeled. Modeling renewable energy is particularly relevant whenever 'credits' are given in LCAs or product carbon footprints (PCFs) for the use of renewable energies, as unjustified credits may lead to wrong conclusions and recommendations. For instance, a company that produces solar power at its factory and feeds it into the grid to get the renewable energy subsidies cannot get an additional low carbon electricity credit, because the renewable energy benefit is socialized and part of the grid. The situation in Norway can serve as an example for the consumption/production mix issue, because its electricity production mix contains a high amount of renewable electricity, but the consumption mix often contains relevant shares of imported energy from fossil sources due to high export sales of renewable energy (Ekvall 2002; Curran et al. 2001). This leads to an underestimation of e.g. the global warming potential for goods produced in Norway when the Norwegian electricity production mix is used.

The existing accounting standards and frameworks like ISO 14044 (2006), ISO/TS 14067 (2013) or GHG Protocol (GHG 2011) just address that double counting has to be avoided, but no practical guidance is given on how a consistent yet practical modeling approach could work. Some authors proposed criteria for allowing credits for renewable energy, but these criteria either do not fully solve the double-counting issue or are too restrictive for getting 'credits' for the use of renewable energy at all (Grießhammer and Hochfeld 2009).

To deal with renewable energy in LCA modeling in a robust way, a consistent shift in the relevant background databases from electricity production mixes to the electricity consumption mixes including the trade of electricity is needed. In addition, more specific accounting guidelines for modeling renewable energy are required, which avoids double counting without being too restrictive in application.

3.1.3 Biogenic Carbon

Biogenic carbon refers to carbon dioxide stored in or released from biomass. The substitution of fossil carbon with biogenic carbon is one potential solution to abate climate change. The substitution can occur in form of biofuels, renewable fibers in plastics or bio-based chemicals and polymers. In addition, biogenic carbon naturally occurs in wood products or in the pulp and paper sectors (Finkbeiner et al. 2012).

Currently, there is no common practice of accounting biogenic carbon within LCA. Two distinct ways exist how CO_2 flows associated with biogenic carbon are modeled. A simplified approach excludes all biogenic CO_2 flows from the calculation by assuming that emissions arising directly from biogenic carbon sources are 'carbon-neutral'. The alternative approach specifically models and accounts for biogenic carbon removal from the atmosphere during plant growth of, e.g., a tree and the future CO_2 release of the wood product when disposed (Rabl et al. 2007; van der Voet et al. 2010). Another challenge arises from the modeling of recycled biogenic carbon. The existing LCA literature so far lacks a specific approach to include the recycling of biogenic carbon stored in products, especially with regard to the distribution of greenhouse gas (GHG) removals from the atmosphere. Due to the lack of a consistent approach for the assessment of biogenic carbon, inconsistencies and errors occur frequently within current case studies. This is particularly relevant for product systems containing renewable raw materials such as wood or cellulose-based polymers—especially when secondary materials or recycling are involved.

Several authors (Luo et al. 2009; Rabl et al. 2007) and even industry associations (e.g. Biotechnology Industry Organization) have already argued against the simple exclusion of biogenic carbon from the LCA models and have proposed to include them as individual flows at each stage of the inventory. Some of the reasons given include the better transparency, the fact that biogenic carbon can be transformed into flows other than CO_2 (e.g. CH_4) or the fact that biogenic carbon can, according to Luo et al. (2009), 'escape' co-product allocations (if the biogenic carbon is not modeled explicitly, it cannot be considered in allocation procedures).

To address the challenge to model biogenic carbon properly, it was proposed by Finkbeiner et al. (2012) to do an explicit accounting of inputs (GHG removals) and outputs (GHG releases) of biogenic carbon flows instead of assuming carbon neutrality *per se*. To deal with the recycling challenge, the same allocation principles have to apply consistently for both burdens (GHG releases) and benefits (GHG removals). That means if burdens are shared between life cycles, also benefits need to be shared between them (see Sect. 3.1.6 on allocation). However, even though a procedure was proposed for modeling biogenic carbon, the approach still allows for different solutions, and further enhancement and harmonization is needed.

3.1.4 Delayed Emissions

Delayed emissions are emissions that are released to the environment with a time delay. The issue of considering delayed emissions in LCA was brought up during the standardization and methodological discussions on carbon footprinting, one impact category within LCA. Therefore, delayed emissions are addressed here for the example of CO_2, even though similar considerations apply for other substances as well.

However, so far, modeling of delayed emissions for products containing, for example, biogenic carbon is inconsistent within LCA. As a consequence, different approaches lead to different results, e.g. to different carbon footprints of products. Modeling of delayed emissions is especially relevant whenever there are differences in the timing of the emission release and particularly for products with long lifetimes. One approach to model delayed emissions is to provide a methodological incentive for delayed emissions so that (biogenic) carbon stored in products is supposed to get a better footprint if the product keeps the carbon longer in the technosphere and delays its release back to the atmosphere. The first version of the British carbon footprint specification PAS 2050 (2008) introduced such a method that uses a discounting approach for delayed carbon emissions. The approach proposed gave delayed emissions a lower weight than emissions that occur immediately, and emissions occurring after 100 years were not considered at all. The same discounting approach was proposed for products with a life cycle of more than 10 years. However, such an approach represents a methodological inconsistency with the ISO-standards of LCA, because no time cut-off for delayed emissions is considered there (ISO 14040 2006; ISO 14044 2006). In addition, such an approach leads to many case study artifacts. For many materials like plastics, the most environmentally preferable end-of-life-treatment option would be a landfill, because there are hardly any emissions occurring within 100 years. Therefore, the more recent carbon footprint standards (ISO/TS 14067 2013; WRI/WBCSD 2004) do not allow a discounting of delayed emissions. They just offer the option to report them separately. Even the PAS 2050 stepped back during its recent revisions and does not allow subtracting delayed emissions from the total emissions anymore (PAS 2050 2011). However, in some studies delayed emissions are still considered due to a lack of awareness or a misinterpretation of the standards.

The solution to this inconsistency is to promote the use of the proper LCA standards as constitution of LCA (Finkbeiner et al. 2012) which do not allow discounting of delayed emissions.

3.1.5 Improbable Events

An improbable event is understood as an event which is unexpected and not likely to happen (OED 2013). It can lead to both positive and negative impacts. The potential impacts of such an event can be evaluated by means of risk assessment (referring to probability and magnitude of the event) (ISO 31000 2009).

Within LCA, currently only steady-state or standard operations are considered. Improbable events—deviations from standard operations or procedures (e.g. an additional extensive cleaning step in a process, a non-routine exchange of a catalyst or machine or an accident)—are not assessed. Hence, improbable events are not covered within the inventory of LCA studies, and their potential impacts are neglected within the impact assessment results. Improbable events (e.g. nuclear meltdown) can significantly influence elementary flows (quantitatively and/or qualitatively) and consequentially affect overall LCA results. For a comprehensive and realistic assessment of potential environmental impacts of e.g. technologies, improbable events need to be taken into account within LCA. Exclusion of the effects of improbable events could lead to wrong conclusions. The degree of impact caused by an improbable event depends on the severity of the event and the type and number of affected elementary flows.

Currently in LCA, risks are considered only within impact assessment (Tukker 2002) where generic risk assessment approaches are used to model human toxicity potential or ecotoxicity potential (Nishioka et al. 2002; Landsiedel and Saling 2002). On the inventory level, the consideration of additional elementary flows caused by improbable events is missing. The resulting risks are not included in current LCA studies.

Improbable events need to be defined in the goal and scope phase and included in the LCI data to account for all potential environmental impacts associated with products and processes. This could be done by means of scenario analyses, e.g. by considering worst case scenarios. However, it may be difficult to quantify and model the full scope of an improbable event. Furthermore, a respective probability range regarding the deviation from the steady-state/standard procedure needs to be considered for the assessment of related impacts.

3.1.6 Allocation

Allocation is defined as "partitioning the input or output flows of a process or a product system between the product system under study and one or more other product systems" (ISO 14040 2006; ISO 14044 2006). An allocation problem arises due to multifunctional processes as the contribution of individual products/processes to the environmental burden is not obvious.

The handling of allocation is clearly a significant concern within LCA studies. Consistent allocation procedures need to be applied for the same multi-functional process (otherwise, unintended ignorance or unintended double-counting of environmental burdens can occur). This is particularly relevant if co-products from one production system are used in different sectors.

To deal with allocation problems, the international standard ISO 14044 (2006) provides a well-known hierarchy of steps. Whenever possible, allocation should be avoided by means of process subdivision or system expansion. If allocation cannot be avoided, physical relationships should be considered and, if not possible, other relationships may be used. Existing allocation procedures for co-products and recycling are presented in the following two subsections.

1. Allocation procedures for co-products
   Allocation in case of co-/by-products is challenging, as the selection of the allocation procedure (e.g. based on mass, calorific value, price, etc.) is not based on science but on value choices. Hence, there is no right or wrong—only a more or less appropriate solution for the specific case.
   As the choice for the allocation procedure often influences the result of an LCA study significantly, the missing scientific basis and consensus between stakeholders on how to handle allocation can be seen as a gap in LCA. This is especially relevant in LCAs of metals derived from ores in companion with other metals (e.g. copper-zinc-gold ores) and animal husbandry (e.g. cows producing milk, leather, meat, and manure) as the choice of the allocation method has a significant influence on the results.
   With process subdivision practitioners can reduce but hardly eliminate all allocation issues as processes may not consist of physically separable sub-processes (very frequent in chemical reactions leading to several substances) (Reap et al. 2008a; Ekvall and Finnveden 2001). Furthermore, system boundary expansion can be applied: Either the functional unit is expanded (e.g. not only milk, but leather, meat, and manure are considered with the milk) or environmental effects of similar products systems are subtracted from the multi-output system

(e.g. environmental effects of gravel production are subtracted from a blast furnace producing steel and slag as a by-product, assuming that slag substitutes gravel in road construction). However, in the first case practitioners may create another allocation problem by including additional processes (Ekvall and Finnveden 2009). In the second case, the risk is that credit is given for a product, which does not reflect the real substitution situation. Both, process subdivision and system boundary expansion, can lead to more data, time, and cost demands, and cause uncertainty (Reap et al. 2008b).
As allocation and associated challenges can neither be avoided nor solved completely, the best way of dealing with it is to apply different allocation solutions and to analyze the results in a sensitivity analysis as suggested respectively required for comparative studies in ISO 14044 (2006).

2. Allocation procedures for recycling
If recycling processes are considered, burdens from primary material production, recycling processes and disposal processes have to be allocated between the life cycles using the material.
The existing requirements in the international standards of LCA, ISO 14040 (2006) and ISO 14044 (2006), are of generic nature. Consequently, no generally accepted approach exists on how to deal with secondary materials recovered from recycling processes. The assessment of recycling processes can have a decisive influence on the overall results of LCA studies and, thus, is relevant for all product systems using or generating secondary materials.
ISO states that "reuse and recycling may change the inherent properties of materials" (e.g. down-cycling in case of plastics recycling (Kuswanti et al. 2003)), and that changes in the inherent properties have to be considered (ISO 14040 2006; ISO 14044 2006). The lack of a clear definition of the term "inherent properties" leads to inconsistencies. Depending on whether inherent properties of materials are changed or not, a distinction can be made between closed- and open-loop recycling. Closed-loop describes the return of material to the same product system (real closed loop) or the return to a different system without changes in the inherent properties of the material (quasi closed loop). Open-loop means, that the material is recycled into a different product system and inherent properties are changed (Ekvall and Tillmann 1997; Klöpffer 1996). In case of closed-loop recycling the allocation of the environmental burden of the primary material production can be avoided, e.g. via system expansion, as the use of secondary material displaces the use of virgin (primary) materials.
The application of existing allocation methods in open-loop recycling systems depends on value judgments, often reflecting the various interests of different stakeholder groups. Thus, the handling of credits and burdens in LCA can lead to under- or overestimation of environmental impacts associated with single life cycles (Reap et al. 2008a).
Several methods of how to account for recycling were developed. Commonly applied are the two extreme approaches of first and last responsibility, also known as recycled content (secondary material source does not carry any burden, because no credit is given for the primary production) and avoided burden

approach (secondary material source carries a burden, because full credit for the primary production burden is given) (Frischknecht 2010; Klöpffer 1996). In addition, several approaches exist in between, e.g. quality-based methods (Azapagic and Clift 1999), economic approaches (Werner and Richter 2000; Guinée et al. 2004) and others (Kim et al. 1997; Wötzel 2007), but no agreement regarding a preferred method has been established so far.

A possible solution seems hard to find, as no allocation method will be applicable in every case (EPA 1993) and different approaches lead to different incentives. Therefore, Curran (2007) recommends to focus on a macroeconomic point of view instead of staying fixed to single processes: to avoid green washing, credits shall only be granted if recycling of the secondary material actually takes place in reality. Neugebauer and Finkbeiner (2012) developed a so called Multi-Recycling-Approach, focusing on the pool of materials, by building up an environmental profile for a certain material equally including the primary and secondary production route. Additionally an evaluation scheme is proposed to assess conservation respectively changes in the inherent material properties (Neugebauer and Finkbeiner 2012).

For consistent modeling of allocation in case of recycling, a consensus for material/product group specific allocation procedures is needed. Until this is achieved, sensitivity analysis should be performed to reflect the influence of the chosen allocation procedure on the results.

3.1.7 Functional Unit

The functional unit is defined as “quantified performance of a product system for use as a reference unit” with the purpose “to provide a reference to which the inputs and outputs are related [and] to ensure comparability of LCA results” (ISO 14040 2006; ISO 14044 2006). According to the International Reference Life Cycle Data system (ILCD) Handbook (EC-JRC 2010b), the functional unit should be defined along the question: ‘what’, ‘how much’, ‘how well’, and ‘for how long’ to support valid comparisons between products.

Due to changing consumption patterns, complex economic systems, and products with multiple functions, the selection of a functional unit is a challenge since different functional units lead to different results for the same product system. This is relevant as a restriction to a strict, functional equivalent may not reflect the reality very well (Hischier and Reichart 2003; Reap et al. 2003; Reap et al. 2008a; Cooper 2003). Realistic modeling is also challenged by issues like lifetime, performance, system dependency, and handling of non-quantifiable or difficult-to-quantify functions (Cooper 2003), e.g. aesthetics of a product. Such information might get lost when defining the functional unit. Moreover, a reasonable straightforward relation of impacts along the life cycle to a functional unit can be questionable, especially when functions are difficult to quantify and effects are included that may be space, time and threshold dependent (ISO 14044 2006; ANEC 2012).

As parts of environmental aspects may not be treated properly when comparing multifunctional products based on one functional unit, Hischier and Reichart (2003) recommend to apply several approaches to consider the multifunctionality—and thus, to better reflect the reality like e.g. ISO/TR 14049 (2012). With its concept of user acceptance (Lagerstedt et al. 2003), ISO/TR 14049 (2012) provides a possibility to compare multifunctional products which are still considered equivalent by the users. Cooper (2003) suggests specifying the functional unit for comparative analyses, e.g. by differentiating between the functions of systems and subsystems and using different functional units when needed.

The challenges described above refer to the relative approach of LCA as such as the reference to a functional unit is the basis of any LCA study. As a consequence, no clear solution exists for this inherent challenge.

3.2 Impact Assessment Aspects

This section analyzes gaps and challenges regarding impact assessment aspects. Following the established structure of areas of protection, the impact assessment aspects are further differentiated into 'human health' (Sect. 3.2.1), 'ecosystem' (Sect. 3.2.2), and 'resources' (Sect. 3.2.3).

3.2.1 Human Health

This section analyzes gaps and challenges regarding impact assessment methods referring to the area of protection 'human health'. It focuses on the impact category 'human toxicity' and related challenges like 'completeness and cumulative effects' or inclusion of 'direct health effects', 'particulate matter', 'nanomaterials', 'endocrine disruptors' and 'microbiological pollution'. Moreover, challenges related to the impact categories 'noise' and 'odor' are discussed.

Since the concept of SLCA emerged, there is some debate whether human health issues should be treated as a social or as an environmental aspect. However, human health has traditionally been assessed within LCA and many studies and models are available. Thus, gaps and challenges related to human health aspects are included in this analysis of (environmental) LCA.

A simplified overview of potential impacts on 'human health' which are currently insufficiently covered in LCA is shown in Fig. 7.3.

Human Toxicity

Health effects of toxic substances are traditionally covered in LCA and either addressed at midpoint level or aggregated at endpoint level (damage to human health). Currently, the consensus model is USEtox ('USE' in the acronym stands for the

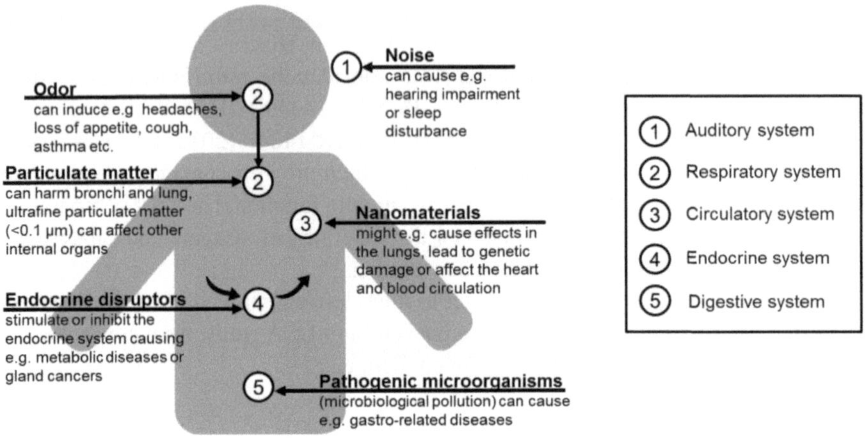

Fig. 7.3 Schematic overview of potential impacts on human health caused by toxic substances, odor and noise

UNEP/SETAC life cycle initiative) (Henderson et al. 2011; Rosenbaum et al. 2008; Rosenbaum et al. 2011) providing guidance for modeling health effects in LCA based on human exposure and toxicity. However, the discussion of human toxicity assessment is far from being solved. General challenges encompass the absence of regionalized and inventory dependent characterization factors and lacking consistency in fate, exposure and effect evaluation. Beyond that, available models are not complete regarding the chemicals which are potentially relevant for human toxicity and neglect cumulative effects of chemicals. This leads to data asymmetry in comparative LCAs. Further challenges related to human toxicity encompass the modeling of 'direct health effects', 'particulate matter', 'nanomaterials', 'endocrine disruptors', and 'microbiological pollution'.

These challenges are addressed in the following sub-sections. As they can be referred to ecosystem as well, they are briefly taken up again in section 'ecotoxicity' (see Sect. 3.2.2.1).

Completeness and cumulative effects. Multitudinous chemicals can cause toxic effects on humans. However, characterization models currently used in LCIA cover only a small fraction of the potentially toxic chemicals. In addition, toxicity is analyzed for individual substances only, and cumulative toxic effects due to exposure to a combination of substances, potentially increasing the toxic effect, cannot be assessed so far. As a consequence, potential toxicity impacts are underrepresented in LCAs of products systems from, e.g., the pharmaceutical sector. Concerning the incompleteness of toxic chemicals covered within LCA, the number of characterization factors has already been increasing steadily from a few hundred in Guinée et al. (2002) to more than 3,000 in the current USEtox model (Rosenbaum et al. 2008). Within the on-going research project 'LC-IMPACT' further methods and characterization factors for ecotoxicity and human toxicity are developed (Rosenbaum et al. 2012). The consensus distribution model (multimedia fate model) has been developed for organic chemicals, not for salts, surfactants, acids and bases. For a recent survey of models see Klöpffer (2012).

Despite this progress, even more characterization factors for toxic chemicals should be determined and potential interrelations should be assessed to achieve a more complete coverage (especially regarding chemicals with potentially high toxicity and/or relatively high emission levels). Furthermore, current characterization models need to assess 'unconventional' toxic effects caused by substances like nanomaterials or endocrine disruptors (see respective sections). The incompleteness of existing LCIAs is a problem of lacking characterization factors and undefined fate models/impact pathways. Additionally, the assessment of cumulative toxic effects poses a substantial challenge on the inventory level. To assess cumulative effects, inventories need to give insight in which of the chemicals are likely to be emitted together, and inventory dependent characterization factors should be developed for the cumulative effects of these chemicals.

Direct health effects. Direct health effects can occur if humans are exposed directly to toxic substances via the respiratory system, gastrointestinal tract, mucosa or skin. Contrary to existing characterization models for human toxicity that comprise default environmental fate models, direct health effects should be accounted for without prior dilution or decay.

Currently, direct health effects are not considered within LCA. It should be noted that LCA was not developed and is not intended to study such effects in detail. This is the domain of tools like risk assessment. However, neglecting direct impact pathways completely can lead to an underestimation of potential toxic effects of, e.g., paint or toys emitting toxic chemicals (Becker et al. 2010b; Guney and Zagury 2011). The inclusion of some impact pathways that consider direct health effects improves the LCIA—acknowledging that these remain potential impacts below the level of risk assessment results.

Several studies evaluate direct impacts of products on humans. These include possible direct health effects of food intake (Juraske et al. 2009), pacifiers and shampoo (Henderson et al. 2012), flame retardants in impregnated textiles and the assessment of indoor emission of building materials for dwellings (Meijer 2007) or chairs (Skaar and Jørgensen 2013). A new version of USEtox has been announced as an outcome of the LC-IMPACT project, enabling the assessment of indoor emissions by considering average room volumes of houses in an indoor fate model (Ernstoff 2012). This latest developments show a promising way to address (at least some) direct health effects in LCA.

The inclusion of direct health effects leads to additional requirements on the resolution of the inventory, as direct exposure has to be separated from releases to the environment. While effect factors can be used from existing characterization models, the fate models and intake fractions might have to be adjusted.

Particulate matter. Particulate matter (PM) refers to fine particles below 10 µm of particle size, forming aerosols. PM can be either man-made (e.g. from combustion in vehicles and plants) or of natural origin and can have a strong impact on human health (Brunekreef and Holgate 2002). Particles smaller than 10 µm (PM_{10}) can enter into the bronchi and lungs, particles smaller than 2.5 µm ($PM_{2.5}$) tend to penetrate into the gas exchange regions of the lung (Delfino et al. 2005), and so called ultrafine particles (below 0.1 μm) may even pass through the lungs and cause health effects in other organs (Sioutas et al. 2005).

Even though different LCIA models addressing PM are available (Greco et al. 2007; Spadaro 2004; van Zelm et al. 2008; Humbert et al. 2011), consistency in fate, exposure and effect evaluation (Potting et al. 2007) are lacking and a more comprehensive approach is needed. In most models a linear, no-threshold dose-response relationship is assumed, although Pope et al. (2009) demonstrated a log-linear relationship. Moreover, toxic effects of ultrafine particles and chemicals attached to particulate matter are currently neglected. LCIA models cannot replace detailed risk assessment analysis for these issues, but the simplified: the inclusion of the potential PM impacts in LCA is still relevant, e.g. in the power and heat sector. New combustion technologies are implemented to reduce the carbon emissions but could also result in different PM emissions (Koornneef et al. 2010) and/or changing particle-size distribution. Thus, trade-offs may occur between the intended environmental benefits of the technology and the potential human health damages due to (changed) PM emissions.

Up to now, only a few of the challenges of existing models are tackled, e.g. some research groups try to set up internally consistent intake fraction values (Humbert et al. 2011).

Additional epidemiological studies of $PM_{2.5}$ and below will help to make the assessment more robust. Chemicals attached to particles have to be included in the inventory, and their toxic effects need to be analyzed in more detail. Finally, case studies are needed to examine data availability to test existing methods and to inspire new methodological developments for a more comprehensive assessment of PM within LCA.

Nanomaterials. Nanomaterials are manufactured/engineered materials with at least one dimension below 100 nm (Gavankar et al. 2012; EC 2012c). Nanomaterials provide increased strength, chemical reactivity or conductivity and are used in many different sectors with growing production volumes in recent years. Nanomaterials exhibit unique behavior depending not only on chemical composition, structure and shape, but also on interaction with organisms and other pollutants in different environmental media (Gavankar et al. 2012; EC 2012c). According to Birnbaum and Jung (2011) they may have relevant effects on human health, as unintended exposure can result in their presence within the body, with unknown biological consequences, e.g. contribution to lung cancer (Becker et al. 2010a) and other effects on the lungs, brain and blood circulation (EU 2009).

Currently, potential impacts of nanomaterials are not included in LCA case studies, as there are gaps on both inventory and impact assessment level. Challenges on the inventory side encompass, for example, missing data of the specific production processes of nanomaterials and the emissions of the nanomaterials itself (Som et al. 2010; Seager and Linkov 2008). Regarding the impact assessment, different cause-effect chains have to be analyzed for different nanomaterials. However, toxicological characterization models for nanomaterials do not yet exist, and as a consequence no characterization factors are available (Gavankar et al. 2012; Som et al. 2010).

The consideration of the potential effects of nanomaterials is relevant and could have a significant influence on the results, in particular with regard to LCAs for sectors using a wide range of nanomaterials, e.g. cosmetics, electronics, textiles, etc. (EC 2012c).

Only a few LCA studies modeled the release of certain nanomaterials (e.g TiN, TiAlN, Ti + TiAlN, carbon nanotubes and others) (Gottschalk et al. 2009; Gavankar et al. 2012; Müller and Nowack 2008), but they remain on the inventory level and are not representative for the multitude of different nanomaterials.

Results from the OECD working group on nanomaterials, e.g. lists of manufactured nanomaterials, exposure measurements or information on safety evaluation and risk assessment (OECD 2013), could be used as a starting point both on the inventory and impact level. The effects of nanomaterials on human health could be included within the existing impact categories, such as human toxicity.

Endocrine disruptors. Endocrine disruptors are hormonally active substances released into the environment that can interact or interfere with hormonal activity (EFSA 2013; EC 2012b). They stimulate or inhibit the endocrine system. Human health can be affected by reproductive development/system disorders, metabolic issues, cardiovascular diseases and gland cancers (US EPA 2012; EEA 2012).

So far, LCA does not fully account for impacts of endocrine disruptors. The information of endocrine disruptors is still limited, and the corresponding causes and effects are complex. Currently, only about 70 substances are listed as main endocrine disruptors and studied internationally (US EPA 2012; EEA 2012). The resulting impacts cannot be taken quantitatively into account in LCIA, as there is currently no epidemiological framework available that covers bioaccumulation, multiple causality, latency, and low doses (US Green Building Council 2008). The assessment of potential effects of the release of endocrine disruptors into the environment is, however, relevant in LCA studies of, for example, pharmacy products, plastics, consumer products or pesticides (Frischknecht et al. 2009), as endocrine disruptors pose a significant concern for human health (Diamanti-Kandarakis et al. 2009).

Some methods in LCA propose an evaluation of the endocrine disruptors in surface or saltwater; for example, the Ecological Scarcity Method. This method uses the estrogenic potential (kg E2-eq/kg) for calculating eco-factors based on yeast estrogenic screening, an accepted method in this research field (Frischknecht et al. 2009). However, since different chemicals can vary widely in their persistence and potency, mass totals are a very crude indication of comparative risk between products.

Due to the novelty of the gap, more inventory and characterization data for endocrine disruptors are required as a starting point for developing a comprehensive solution.

Microbiological pollution. Microbiological pollution refers mainly to the pollution of freshwater by pathogenic microorganisms. Infectious microorganisms can be grouped into bacteria, viruses, protozoa, and helminthes (NRMMC 2006) and cause mainly gastro-related diseases.

In LCA, there are currently neither elementary flows defined on inventory level, nor impact categories available on impact assessment level. Microbiological pollution is particularly relevant for systems characterized by wet conditions, temperature changes, and availability of organic matter. Hence, it should be considered, for example, in LCAs of waste water treatment plants, water dispensers or biogas plants.

To address microbiological pollution, Larsen et al. (2009) proposed a first framework to assess emission of pathogens into recreational water bodies and unintended swallowing of water during bathing measured in disability adjusted life years (DALYs). This framework follows the structure of risk assessment including hazard identification, dose-response analysis, exposure assessment, and risk characterization. With regard to LCA, Larsen et al. (2009) noted that models for dose response presented in NRMMC (2006) may be usable for a simplified assessment in LCA.

In order to address impacts of microbiological pollution on human health in LCA, adequate elementary flows have to be defined providing information on the number and kind of pathogens emitted. Furthermore, additional impact pathways have to be identified, and characterization models have to be developed to assess impacts resulting from microbiological pollution. Such models should include the fate of pathogenic microorganisms comprising distribution through environmental media and survival rates of pathogens. After that, the uptake of microorganisms into the human body (via different uptake routes) has to be considered, and health effects should be predicted by means of clinical dose-response relations.

Noise

Noise can be regarded as 'unwanted sound' which possibly affects human health, as it may, for example, impair cognitive abilities (Clark et al. 2006), cause sleep disturbance (Griefahn et al. 2006), increase the risk for heart diseases (van Kempen et al. 2002) or lead to hearing impairments.

Although methodologies to assess noise are available and an LCA midpoint category was already suggested by Heijungs et al. (1992) and Guinée et al. (2002), it is so far hardly addressed within LCA case studies. The main reason is the lack of inventory data. Furthermore, the relation of noise of a production process to the functional unit is not straightforward. Neglecting this impact category may lead to incomplete and falsified results in LCAs of transport systems, for example, as potentially significant impacts on human health are disregarded.

Franco et al. (2010) proposed a framework to assess impacts of traffic noise based on the method developed by Müller-Wenk (2002, 2004) where effects of traffic noise are considered, determining the number of annoyed persons attributed to kilometer per vehicles. In addition, the Centre of Environmental Science (CML Leiden, The Netherlands) proposed a method aggregating physical sound levels (Heijungs et al. 1992; Guinée et al. 2002). Further methodologies for assessing noise are available (Althaus et al. 2009; Meijer et al. 2006; Reap et al. 2008b;

Cucurachi et al. 2012), but a consistent framework and comprehensive inventory data are still needed.

To include noise in LCA it is recommended in order to conduct more case studies applying, testing and improving available methods.

Odor

Odor can be regarded as ‘unwanted and unpleasant smell’, caused by volatile chemical compounds. Many effects of odor are covered within other impact categories (e.g. volatile organic compounds—VOC emissions in human toxicity). Furthermore, odor can affect quality of life and human well-being which could be considered outside the scope of LCA. However, as odor can also have direct effects on human health, e.g. induce headache, sleeplessness, loss of appetite, sickness, nervousness, cough or asthma (Blaisdell 2007), it should be considered within the scope of LCA.

So far, odor is rarely analyzed in LCA due to the lack of both inventory data and robust impact assessment methods. Addressing odor in LCA can be relevant especially with regard to e.g. wastewater treatment systems or biogas technologies, as the impacts of odor on humans can be significant and alter the results of LCA studies.

Several approaches have been suggested for the inclusion of odor into LCA. Heijungs et al. (1992) and Jolliet et al. (2004) developed midpoint indicators. Guinée et al. (2002) proposed to inverse the odor threshold values (OTV) for characterizing odor, so that the smell creation potential (SCP) can be described by dividing the emissions of a substance by the OTV value. Marchand et al. (2012) proposed a site dependent approach for odor assessment in waste management, based on analyzing concentrations, fate using UseTOX (Rosenbaum et al. 2008),—and exposure.

Within the existing impact category and proposals, odor concentration is assessed by determining VOC emissions (kg/m^3), but with regard to a complete characterization method additional factors like deposition, evaporation, chemical conversion and meteorological conditions have to be considered. For comprehensive assessment within LCA, inventory data need to be collected and case studies should be performed to develop an applicable impact assessment methodology.

3.2.2 Ecosystem

This section analyzes gaps and challenges regarding impact assessment methods referring to the area of protection ‘ecosystem’. It focuses on the impact categories ‘ecotoxicity’, ‘biodiversity’, ‘biological invasion’, ‘direct non-intended killing of animals’ as well as ‘land use and land-use change’. The impact category ‘noise’, which was already discussed in Sect. 3.2.1 with regard to human health, is briefly taken up within the section ‘ecotoxicity’ with regard to its potential impacts on animals.

Ecotoxicity

In the previous section, methodological gaps concerning the assessment of effects caused by toxic substances on human health have been comprehensively discussed. Obviously many of those challenges and proposed solutions apply for the analysis of ecotoxicity, too. In order to avoid repetition, this section revisits these gaps from an ecotoxicity perspective and considers differences and similarities in the analysis of gaps identified for human toxicity.

Despite the fact that for the assessment of ecotoxicity different fate and effect models have to be used than for human toxicity, challenges regarding 'completeness and cumulative effects of chemicals', 'nanomaterials', and 'endocrine disruptors' are similar (see Sect. 3.2.1). The lack of detailed toxicological knowledge is the main reason for incomplete or missing characterization factors. With regard to endocrine disruptors, derivation of characterization factors for ecotoxic impacts might be easier, as there are laboratory studies available analyzing effects on animals directly while effects on humans can only be observed indirectly (Matthiessen 2000). On the other hand, recent studies on animals showed that the dose-response relation is not always linear and that endocrine disrupting chemicals may have effects at low doses but no effects at high doses (Vandenberg et al. 2012). This is a significant challenge for the development of characterization models.

Particulate matter might affect ecosystem as well as it can harm plant life and thus animals which are dependent on plants as feed. Moreover, animals can also be affected by breathing in the particulates, similar to humans (HCES 2013). However, as emission of fine dust is mainly relevant in urban areas (except forest fires), ecotoxic effects of particulate matter can be regarded as less important.

Microbiological pollution can be relevant for ecosystems, too, as e.g. animals might be affected by pathogenic microorganisms. Considering that only some species are sensitive to pathogenic pollution, potential characterization models for ecotoxicity would have to focus on target species.

Toxicity impacts caused by direct health effects, e.g. indoor emissions, are relevant for human health but generally not for ecosystems as exposure to indoor pollutants or products emitting toxic chemicals directly into an animal's body, which is part of the ecosystems, are rather unlikely.

Generally, odor could also have impacts on ecosystems. As no scientific proof could be found whether animals are affected by odors or not, more basic research is needed in this area before discussing potential impact assessment approaches.

Noise can affect animals and cause auditory damage due to high noise levels or physiological changes, e.g. increased heart rate, problems with respiration, and behavioral changes, e.g. changes in migration patterns (Cornman 2003). Currently, that is not addressed in LCA. Challenges and possible first steps to integrate noise effects on animals in LCA can be considered similar to those discussed in Sect. 3.2.1.2.

Biodiversity

Biodiversity is the "variability among living organisms from all sources including, inter alia, terrestrial, marine and other aquatic ecosystems and the ecological complexes of which they are parts: this includes diversity within species, between species and of the ecosystems" (UN 1992). Biodiversity is important for the stability of habitats and ecosystems and highly sensitive to the alteration of all environmental aspects, e.g. quantitative and qualitative changes in air, water and soil (Gontier et al. 2006).

The already existing concepts to include biodiversity in LCA as midpoint or as endpoint category have been regarded as too limited and vague to be applied, and not equally suitable and valid to all kinds of ecosystems (differences between terrestrial, freshwater and marine needs to be considered) (Noss 1990). Thus, an agreed overall method is still missing within LCA. Over the last decades a strong decline in biological diversity occurred (Curran et al. 2011). Human activities can affect biodiversity directly e.g. by agricultural production leading to land use (Michelsen et al. 2012), transport processes (e.g. ships) causing biological invasion (Schenck 2001), plastic packaging risking littering (Wong et al. 1974). They can also affect biodiversity indirectly, e.g. by pharmaceutical products causing the release of nanomaterials (Som et al. 2010) or endocrine disruptors (Frischknecht et al. 2009). Therefore, the respective impacts need to be considered within LCA. Consideration of effects on biodiversity is relevant e.g. for LCA studies on agricultural and food products, as direct nature occupation occurs in this connection (Knudsen and Halberg 2007).

In LCA, biodiversity besides ecosystem functions is often used to assess ecosystem quality (Henzen 2008) and discussed both within midpoint approaches (Guinée et al. 2002) and endpoint approaches to LCIA (Curran et al. 2001). Curran et al. (2011) and Watson et al. (2005) discuss endpoints of biodiversity in LCA and name five drivers for biodiversity loss. Three of them with regard to habitat change, climate change and pollution are represented in current impacts categories (e.g. land use, eutrophication, and acidification). The other two drivers for biodiversity loss, invasive species and overexploitation of ecosystems, especially overfishing/by-catch (Sects. 3.2.2.4 and 3.2.3.2), are so far hardly represented in any impact category (Curran et al. 2011; Watson et al. 2005). Several indicators (like species richness, variety or number of species, species vulnerability) and approaches for the assessment of biodiversity were proposed (Milà i Canals et al. 2007a; Schenck 2001; Souza et al. 2013; UNEP 2010b), but none of them provides a comprehensive solution to include biodiversity in LCA. Researchers currently analyze the multiple aspects of biodiversity loss on a mostly local level with the help of ecosystem services (UNEP 2010b).

To address this gap, a deeper understanding of the biological, geophysical, and geochemical processes and intensive research is needed (Balmford et al. 2005). Furthermore, already existing indicators, inventory data (Jeffery et al. 2010) and approaches should be revised and extended to achieve regionalized data capturing a representative sample of the diverse terrestrial, freshwater and marine habitats (Curran et al. 2011). Exemplarily, one relevant indicator to describe biodiversity, namely biological invasion, is further discussed in Sect. 3.2.2.3. Another approach is to include biodiversity in existing impact categories within LCA, e.g. combining biodiversity loss with land use (e.g. by including data from the Geographic Information System—GIS in LCA) or global warming (Geyer et al. 2010; Knudsen and Halberg 2007).

Biological Invasion

Biological invasion, or invasion of foreign species into ecosystems, can cause damages in local ecosystems if indigenous species are displaced by invaders.

So far, biological invasion is not considered within LCA. Biological invasion is relevant for processes like transports, especially for shipping: First, channels used for shipping connect ecosystems and allow for a spreading of species. Second, ballast water used to stabilize container ships during empty running is a potential source for biological invasion as alien species, such as jellyfish or mussels, can be transported over long distances and are then released into non-indigenous ecosystems.

In an LCA context, biological invasion is proposed as an indicator describing impact pathways on biodiversity, by means of, e.g., "percent coverage of invasive species within protected areas" (Schenck 2001). However, this can be regarded as an inventory indicator, and a deeper analysis of impacts is missing. Recently, Narščius et al. (2012) published a method which allows for the assessment of biological invasion impacts. The model determines a biological pollution level by relating a classified abundance and spreading range of invasive species to the magnitude of their consequences on communities, habitats and ecosystem functioning.

To include biological invasion in LCA, further research is needed both on an inventory and on an impact assessment level. Regarding the latter, cause-effect mechanisms described in the method proposed by Narščius et al. (2012) might be transferred into a characterization model for LCIA.

Direct Non-Intended Killing of Animals

Direct non-intended killing of animals includes the non-intended mechanical, electrical or thermal causes of direct death of animals by human activities, e.g. impacts due to road traffic, wind or water power plants, oil spills (and other man-caused accidents) or industrial fishing (by-catch). Some of these impacts are related to accidents, but the shredding of fish in water power plants or the by-catch in industrial fishing are not really accidental, they are a non-intended part of standard operation. The term 'direct' is used in this context to differentiate the resulting impacts from the indirect effects on animal health due to other impact categories like ecotoxicity or water use.

The direct non-intended killing of animals is not addressed within LCIA so far, as existing impact categories like ecotoxicity, acidification and eutrophication

(Cleuvers 2003; Goedkoop et al. 2009; Jolliet et al. 2003; Rosenbaum et al. 2008; EPA 2012; Jolliet et al. 2004; EC-JRC 2010a) address potentially lethal impacts on species only indirectly, e.g. in the endpoint models for eutrophication or ecotoxicity. The inclusion of the direct non-intended killing of animals and possible consequences like injury, mortality and even declines in species seem particularly relevant for the following examples: fish mortality due to hydropower plants, mortality of bats due to collision with offshore oil and gas platforms and mortality and injuries of birds caused by electrocutions from power lines (Banks 1979; BPA 2013; NYSERDA 2009; Watts 2010) or by collision with turbine rotors of wind farms (Cole 2011; Ferrer et al. 2012). In addition, direct non-intended killing of animals can also be caused by littering, by-catch (e.g. fish, dolphins), oil spills (and other man-caused accidents) (e.g. affecting birds), and human constructions like swimming pools and building pits (e.g. affecting hedgehogs) (Ambrose et al. 2005; Burger 1993; Erickson et al. 2005; Fleet et al. 2009; Garrity and Levings 1993; Pro Igel 2012).

The direct non-intended killing of animals is already addressed by risk assessment and environmental impact assessment (EIA). Within risk assessment models for example, killing of birds during the operation phase of wind power plants has been displayed to predict, assess and possibly reduce bird mortality (Ferrer et al. 2012). Furthermore, EIA studies exist, which evaluate the killing of birds and bats in connection with the construction of wind farms (Arnett et al. 2007; Smallwood et al. 2007). Although the direct mortality rates vary widely, depending, for example, on the location, the risk for birds and bats can be seen as proven (Cole 2011; Ferrer et al. 2012; Kuvlesky et al. 2007; Smallwood et al. 2007). Besides that, mortality caused by other human constructions is hardly addressed within case studies so far.

To include direct non-intended killing of animals in LCA models, inventory data and indicators need to be developed. As a possible starting point the existing risk and EIA studies can be used. The developed EIA indicators, like total discounted bird years (Cole 2011), can be used, improved and translated into LCA, e.g. inclusion of animals and/or species mortality by means of years of animal life lost (YOALL). Within the development process of indicators, also threshold values shall be included that consider endangered species or specific limitations to avoid population damages.

Land Use and Land Use Change

Since clear definitions of land use (LU) and land use change (LUC) cannot be found in literature, a clear distinction between both is not a straightforward task. For some authors, occupation and transformation are equally responsible for land use and land use change (EC-JRC 2010a; Saad et al. 2011); others consider land use to be related to occupation and land-use change to transformation (Mattila et al. 2011). Some authors consider only biodiversity when assessing land use (Vogtländer et al. 2004), others consider also changes in soil quality, (Milà i Canals et al. 2007b) changes in GHG emissions (Anderson-Teixeira et al. 2011) or the effects on fresh water (EC-JRC 2010a; Saad et al. 2011). Within this section, the two terms are discussed separately based on their development within the LCA community for the last few years. We use Milà i Canals (2007) definition for LU as damage to ecosystems due to the effects of land occupation (of a certain area during a certain time) and transformation (of a certain area). On the other hand, LUC is a general term for the alteration of one land use category to another (Mattila et al. 2012). Assessing LU and LUC is especially important when considering the production of land-intensive products from agriculture, mining or infrastructure.

Land Use The main consequence of land use is the loss of biodiversity (Millennium Ecosystem Assessment 2005). In addition, impacts on ecological functions like changes in soil quality and ecosystem services like biomass production occur (Koellner et al. 2013).

Currently, the impact category land use is hardly applied in LCA. One reason is the lack of a broadly agreed impact assessment method (Mattila et al. 2012). Most of the existing methods are only adding up square meters (in general or based on hemeroby classes), which do not adequately reflect losses of biodiversity or soil quality aspects (Milà i Canals 2007). Lack of sufficient inventory data is another problem as for most of the existing methods, spatial inventory data is needed. Even though background data on square meter level is partly existing, databases are far from being complete (Milà i Canals et al. 2007a; Vogtländer et al. 2004; Mattila et al. 2012).

The ILCD Handbook (EC-JRC 2011) recommends to assess land use with the midpoint indicator soil organic matter (Milà i Canals 2003), which can only reflect some changes in soil quality, but not biodiversity as such (Milà i Canals et al. 2007b). Assessing land use at endpoint level is suggested in Koellner and Scholz (2008) and Goedkoop et al. (2009).

Koellner et al. (2013) introduce guidelines to develop a land use impact assessment for biodiversity and ecosystem services and also recommend specific methods and characterization factors for some of these aspects.

Following Mattila et al. (2012), further steps could be the application of existing midpoint- and endpoint-methods to test their feasibility. As a crucial prerequisite for broader application, inventory data needs to be collected.

Land Use Change LUC is often divided into direct LUC (dLUC) and indirect LUC (iLUC). dLUC occurs when new agricultural land is taken for production and feedstock purposes and therefore, displaces e.g. a forest (Sanchez et al. 2012). iLUC occurs when land currently used for feed or food crops is changed to the production of a different product and the demand for the previous land use remains (Piemonte and Gironi 2011).

So far, no globally accepted and used methodology to assess LUC exists in LCA (Mattila et al. 2012). A challenge in LUC when creating an LCA model is the precise definition of the system. It is tricky to assume whether the burdens will occur immediately or over multiple years of production. Almost every LCA study assumes

LUC will occur immediately. However, it is expected that in reality LUC effects occur over many years (especially for iLUC) (Anderson-Teixeira et al. 2011). Another important challenge that occurs when dealing with land transformation is the estimation of indirect emissions. There is currently no consensus on how this is to be performed in LCA/carbon footprint (CF) studies (Finkbeiner 2013). Considering LUC in LCA is relevant as LUC is appearing to be one of the challenges in LCAs of renewable raw materials and biofuels, "often determining whether or not a biofuel meets GHG reduction thresholds" (Anderson-Teixeira et al. 2011).

Various methods have been proposed to include LUC in LCA and show "extremely different results" (Flysjo et al. 2012). Thus, the iLUC modeling process is not yet mature or sufficiently advanced to be used as policy-making tool (Plevin et al. 2010). Several other studies observe different options to lower the uncertainty when evaluating LUC emissions (Anderson-Teixeira et al. 2011), especially the indirect ones (Sanchez et al. 2012; Jannick H Schmidt 2012). There is also the fundamental issue why indirect market effects are supposed to be included in LUC at all, as no other indirect market effects are included in LCA. In that sense, iLUC is methodologically inconsistent with the basic principles of LCA. It might be consistent with the approach of consequential LCA, but fully depends on uncertain market predictions.

LCA community needs a deeper analysis and discussion, if and how to estimate LUC emissions (Flysjo et al. 2012). If further research achieves a better understanding of LUC, scientifically robust and consistent LUC quantification, factors might be reconsidered as a potential element to be included in LCA and CF. Until then, LUC should be reported along with LCIA results.

3.2.3 Resources

This section discusses gaps and challenges in impact assessment methods referring to the area of protection 'resources'. Besides 'abiotic resources' and 'biotic resources', challenges in impact assessment schemes for 'changes in soil quality', 'desertification' and 'salinization' are analyzed.

Abiotic Resources

Abiotic resources comprise all non-living resources such as minerals, fossil fuels, water and are a relevant input to many products and production processes.

By extracting resources, the concentration in the earth's crust is changed. However, if and to what extent biogeochemical cycles are affected or changed and if environmental changes occur due to resource depletion is not clear. Thus, resource scarcity is not always seen as a true 'environmental impact' (UNEP 2010a). Still, existing LCA characterization models, like abiotic depletion potential (Guinée et al. 2002) or surplus energy (Goedkoop and Spriensma 2001), aim at assessing the environmental dimension of resource depletion. As these models usually consider geologic abundance of resources, they are in practice often used to analyze resource availability for production processes. Yet, when analyzing resource scarcity, an assessment of geologic availability is not enough; limiting socio-economic factors such as country concentration of reserves or monopolistic trade structures need to be taken into account. This is particularly relevant for, e.g., materials needed in future technologies, which are commonly perceived as scarce such as rare earth metals. So far, within LCA studies, these materials do not contribute to geologic depletion in a noticeable manner. Existing indicators deliver no decision support and can even lead to wrong conclusions concerning the actual availability of resources.

A method for considering anthropogenic material stocks in addition to geologic resources was developed (Schneider et al. 2011a) reflecting physical resource availability more realistically. However, for a comprehensive evaluation resource assessment has to go beyond physical considerations. In terms of socio-economic resource availability, several frameworks have been developed (Graedel et al. 2012; EC 2010; Erdmann and Graedel 2011; National Research Council 2008), but only limited research is available regarding an integration or combination with LCA.

To promote a more comprehensive and realistic assessment of resource use and availability within LCA, a method for resource scarcity analysis in LCA, which transfers the socio-economic indicators into impact categories including characterization models is proposed (Schneider et al. 2011b; Schneider et al. 2012; Schneider et al. 2013).

Biotic Resources

Biotic resources, like natural forest or fish, can reproduce and are regarded as living (Guinée et al. 2002). Agricultural products, such as crops, farm animals (including fish-farms) or grown wood, are not considered as biotic resources, but as products derived from resources such as land, water, solar energy, and nutrients.

While impacts associated with the use of biotic resources like land use change, water consumption or global warming are addressed in the respective impact categories, the depletion of biotic resources is often neglected. Depletion of biotic resources occurs if their use exceeds their renewability rates. Hence, it should be assessed in LCA to avoid incomplete assessments of, e.g., fishery or products containing tropical wood.

A potential solution to assess biotic resource depletion, considering stocks in relation to production and renewability rates, is given by Guinée et al. (2002).

However, similar to abiotic resources (see Sect. 3.2.3.1), the availability of biotic resources can be limited by socio-economic factors as well (Schneider et al. 2011b; Schneider et al. 2012). Additionally, factors like vulnerability to natural disasters (e.g. forest fires or pest infestation) and logistic constraints, like storage stability, may affect biotic resource availability (VDI 2013). Consequently, such aspects need to be analyzed in more detail and transferred into characterization models leading to a comprehensive physical and socio-economic assessment of biotic resources.

Changes in Soil Quality

Existing soil quality definitions can be grouped in two main categories: (a) definitions which focus on ecological services like filter, buffer or storage, and (b) definitions which focus on soil uses in terms of soil fertility such as agricultural production. Soil quality might be influenced by soil texture (particle-size distribution), soil structure (framework and position of soil particles), pH value, and presence and availability of nutrients, weeds, pathogens, salts, organic matter, and contaminants (Cowell and Clift 2000).

Even though a number of studies and approaches assess soil quality in LCA, there are still gaps both on inventory and on impact assessment level. So far, no impact assessment method exists which combines numerous interrelated soil characteristics. Furthermore, agricultural practices are influencing soil quality and are not yet considered within LCA (Garrigues et al. 2012). The assessment of soil quality within LCA is relevant for products and services affecting soil quality or use soil, e.g. agriculture, forestry or civil engineering.

Potential impacts caused by heavy metals and pesticides are already considered in the impact category 'terrestrial toxicity'. However, further impact categories on soil quality are still under development, for example soil compaction (Garrigues et al. 2013) and soil erosion (Núnez et al. 2012). For the calculation of inventory data for soil erosion, Wischmeier and Smith (1978) proposed the 'Universal Soil Loss Equation' (USLE); for the soil erosion impact assessment, Bindraban et al. (2012) and Garrigues et al. (2012) suggest using the characterization factors from the 'International Soil Reference and Information Centre' (ISRIC). Mila i Canals (2003) proposed to use soil organic matter as an indicator for soil quality. However, these approaches only consider single soil properties and therefore are not suitable to depict the whole complexity of soil quality. Some approaches consider multiple criteria: Cowell and Clift (2000) combine soil erosion, change in organic matter and soil compaction and the 'Swiss agriculture life cycle assessment—soil quality' (SALCA-SQ) method contains nine physical, biological and chemical properties for Swiss conditions (Garrigues et al. 2012).

To derive a single method for soil quality impact assessment, robust and globally valid impact indicators for changes in soil parameters need to be developed and subsequently combined (Garrigues et al. 2012). As a large number of parameters contribute to soil quality, the development of a single soil quality impact method is challenging. Modifications of even a few soil parameters could have severe consequences in certain regions. Thus, it is suggested to assess desertification and salinization separately, as these processes dominate impacts on soil quality for certain regions in the world. Therefore, methods to assess desertification and salinization are described in separate sections (see Sects. 3.2.3.4 and 3.2.3.5) while other factors like erosion and organic matter are summarized under soil quality. The relation between the impact categories is described in Fig. 7.4.

Desertification

Desertification describes the degradation of land due to drastic changes in soil properties that lead to limitations in soil functions such as the supply of nutrients to plants or the water holding-capacity. Factors like climate variations or human activities (inappropriate farming practices, water management, etc.) might cause desertification. In particular arid, semi-arid and dry sub-humid areas are affected (UN 2011).

Even though desertification is a severe problem, only one approach exists to assess desertification in LCIA (Núñez et al. 2010). Desertification is usually neglected in current LCA studies. Assessing desertification is especially relevant for agricultural products from regions vulnerable to desertification, e.g. cotton produced in Egypt.

The existing methodology by Núñez et al. (2010) is based on a set of four variables (aridity, erosion, aquifer exploitation, and fire risk). Characterization factors for desertification are provided for the main terrestrial ecological regions (Núñez et al. 2010). However, there is limited case study experience to test scientific robustness.

In order to start considering desertification impacts in LCA, the method of Núñez et al. (2010) should be tested in case studies of products providing the risk of desertification. Even though this might be challenging as spatially explicit inventory data is required, case studies are needed to validate and, if necessary, improve the characterization model.

Salinization

Salinization is defined as the accumulation of water-soluble salts in soil or water bodies caused by human activities or natural processes. It may reach to levels that may cause impacts to the environment and humans (Podmore 2009). High levels of salinity can influence the water absorption of plants, may lead to soil degradation by changing the soil texture, affect surface and groundwater properties and could, to some extent, lead to distraction of habitats and biodiversity. Moreover, high salt levels in water can affect infrastructure (e.g. corrosion) (Podmore 2009; Leske and Buckley 2003).

In LCA, salinization is mentioned in relation to land use or freshwater depletion (Jolliet et al. 2004) as well as biodiversity (Amores-Barrero et al. 2013). However, consistent frameworks are missing, and so far, salinization is mostly neglected in LCA case studies (EC-JRC 2010c). Anthropogenic sources for soil salinization are mainly irrigation processes in agriculture. Salinization of water bodies is caused by over-fertilization, (industrial) wastewater, leachate from landfills or by high amounts of road salt (Podmore 2009; Leske and Buckley 2003). Thus, considering salinization is especially relevant for LCAs on agriculture or waste (water) management.

Feitz and Lundie (2002) developed an indicator for soil salinization and irrigated salinity, the so called salinization potential (SP), and proposed a preliminary soil salinization impact model as an indicator for land degradation from poor irrigation practices. Besides soil, Leske and Buckley (2003) also consider salinization of water bodies and proposed a new impact category for salinization including a separate characterization model. The main limitations of these models refer to the restricted

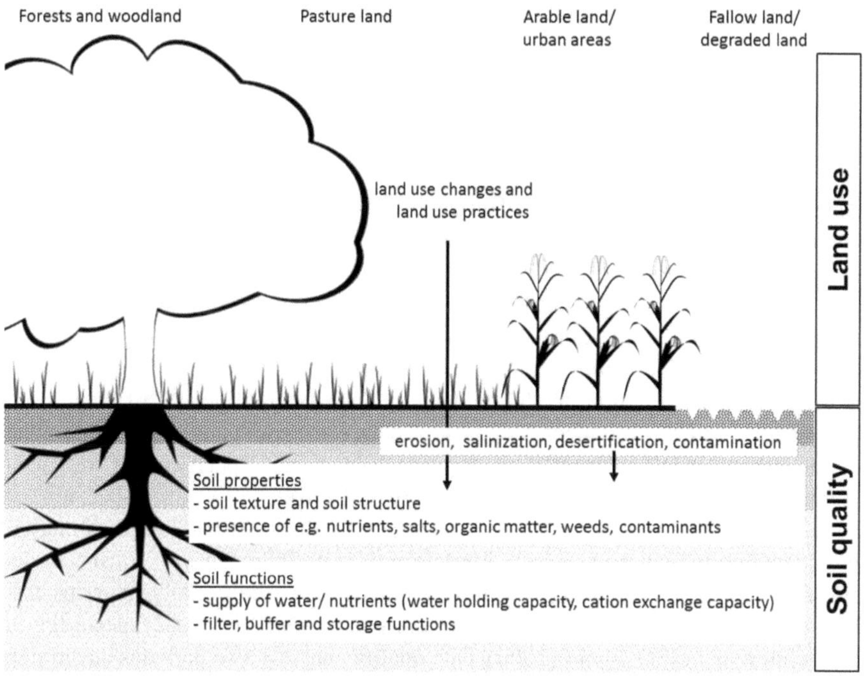

Fig. 7.4 Schematic relation of land use and land use change, changes in soil quality, desertification and salinization

scope—e.g. limitation to irrigation in Feitz and Lundie (2002)—and to the detailed inventory data requirements such as soil composition. Moreover, site specific fate model parameters were so far only developed for Australia (Feitz and Lundie 2002) and South Africa (Leske and Buckley 2004).

To include salinization in LCA, further clarification is needed, whether salinization can be addressed within existing impact categories (e.g. soil quality) as indicated in Fig. 7.4 or a separate midpoint category is required (Jolliet et al. 2004; Reap et al. 2008b). In any case, local or regional salinization potentials would be needed for the assessment of individual regions (Feitz and Lundie 2002; Leske and Buckley 2004).

A simplified overview of the relation between the impact categories 'land use and land use change' discussed in Sect. 3.2.2.5 and 'changes in soil quality', 'desertification' and 'salinization' described in the previous sections is shown in Fig. 7.4.

3.3 Generic Aspects

This section analyzes gaps and challenges referring to overarching aspects of LCA including 'data quality analysis', 'uncertainty analysis' and 'weighting'. Moreover, recent developments aiming at a more comprehensive assessment of potential environmental interventions are discussed. These include 'macroeconomic scale-up' and 'consequential modeling approach' as well as the consideration of 'rebound effects' in LCA.

3.3.1 Data Quality Analysis

According to ISO 14044 (2006), data quality is the "characteristics of data that relate to their ability to satisfy stated requirements" and "should be characterized by both quantitative and qualitative aspects, as well as by the methods used to collect and integrate those data". The requirements for data quality should be specified in the goal and scope and address e.g. information on time-related or geographical coverage, representativeness, sources, data variability as well as information on uncertainty of information, including data, models, assumption (ISO 14044 2006). Data quality can be influenced by a lack of data, wrong and ambiguous data, inaccurate measurements and model assumptions (Baker and Lepech 2009; Ciroth et al. 2004; Heijungs and Huijbregts 2004; Huijbregts 1998).

However, there is a lack of consensus regarding the systematic methodology of how data quality can be assessed. Therefore, it is not adopted in a consistent and constant way in LCA studies (May and Brennan 2003). Additionally, overlaps exist between data quality analysis and uncertainty analysis (Sect. 3.3.2): according to the definitions in ISO 14044 (2006), data quality analysis also includes uncertainty information, and uncertainty analysis also includes data variability. Data quality analysis is relevant for every LCA study, for both background and foreground data, as systematic errors deriving, for example, from wrong data or assumptions, as well as statistical errors from data variability, can significantly influence LCA results. It is particularly important for judging the significance of differences in comparative studies (Canada Mortgage and Housing Corporation 2004; Huijbregts 1998; Notten and Petrie 2003; Sonnemann et al. 2003).

In literature several approaches for data quality analysis exist. Weidema and Wesnaes (1997), for example, introduced a 'pedigree matrix' for a semi-quantitative evaluation of data quality, considering reliability, completeness and temporal as well as geographical correlation. Kennedy et al. (1996) developed a methodology to convert deterministic LCA models into stochastic models to quantify the effects of data quality uncertainty on the result of an LCA. May and Brennan (2003) recommend a quantitative assessment in combination with a separate qualitative assessment on data quality. However, none of these approaches represents a broadly accepted data quality analysis scheme.

Because of the limited robustness of data quality analysis, particular caution has to be paid when drawing conclusions from a study.

3.3.2 Uncertainty Analysis

According to Huijbregts (1998), Sonnemann et al. (2003) and Heijungs and Huijbregts (2004) uncertainties within LCA consist of parameter, scenario and model uncertainties. ISO 14044 (2006) defines uncertainty analysis as a "systematic procedure to quantify the uncertainty introduced in the results of a life cycle inventory analysis due to the cumulative effects of model imprecision, input uncertainty and data variability".

However, terminology is not yet standardized: the terms 'uncertainty', 'variability' and 'sensitivity' are not clearly defined within the LCA community (Heijungs and Huijbregts 2004; Baker and Lepech 2009). Moreover, the relation between data quality analysis (Sect. 3.3.1) and uncertainty analysis is not always clear. The ISO 14044 (2006) definition, for example, includes data variability and moreover describes uncertainty analysis as "additional method for LCIA data quality analysis". Currently, a systematic methodology on how to assess uncertainty is lacking. Uncertainty analysis is relevant in the LCA evaluation/interpretation phase, as uncertainties, e.g. deriving from statistical errors like data variability or model imprecision, can significantly influence the LCA result.

Specific approaches like the Monte Carlo analysis, fuzzy set methods, Bayesien methods or pedigree matrices (Jolliet et al. 2009) are used to determine the uncertainties of the phases within LCA (Sonnemann et al. 2003; Heijungs and Huijbregts 2004; Lloyd and Ries 2007; Lo et al. 2004). But it is still difficult to define, classify and assess the whole range of uncertainties (Benetto and Dujet 2003). Some generic LCI databases like ecoinvent (ecoinvent 2013) already provide information about distributions and data quality to calculate statistical values (Heijungs 2010; Jolliet et al. 2009; Sonnemann et al. 2003). However, systematic errors, like methodological choices concerning cut-off criteria or allocation procedures, are often more relevant than statistical errors due to random scattering of data.

Consequently, the existing uncertainty analysis methods do not really help in defining the significance of results. Furthermore, the need for software and databases which support these approaches is undeniable. To address this challenge, further scientific developments and consensus on proper uncertainty analysis methods are needed to ensure that the uncertainty of uncertainty analysis methods is smaller than the uncertainty of the LCA study itself.

3.3.3 Weighting

According to ISO 14044 (2006), weighting is an optional element which can be used to convert the results of the different impact categories into one single score indicator by using numerical factors based on social, ethical and political value choices from one or more stakeholder groups; "it shall not be used in LCA studies intended to be used in comparative assertions intended to be disclosed to the public" ISO 14044 (ISO 14044 2006).

As different individuals, organizations and societies may have different preferences, it is possible that different parties will reach different weighting results based on the same indicator results or normalized indicator results (ISO 14044 2006), potentially leading to different conclusions. Regularly, LCA results are communicated to LCA nonprofessionals who may prefer single-score results due to clear and easy interpretation. However, whenever single-score indicators are used, existing gaps and shortcomings of weighting methods, which might affect conclusions and tradeoffs between impact categories, would be hidden.

The three most commonly used weighting methods are the panel method, the distance-to-target method and the monetary method. For example, the eco-indicator 99 (Goedkoop and Spriensma 2001) and ReCiPe (Goedkoop et al. 2009) methods, which are based on the panel method, include three different cultural perspectives with different views on nature that can be chosen depending on the involved stakeholders (Huppes and Oers 2011). Those three views are only representative for very few stakeholders and therefore, not generally applicable (Goedkoop et al. 2000). The distance-to-target approach, as applied in the ecoscarcity method (Frischknecht et al. 2009), is based on the current performance of a country in relation to aspirated standards, laws or goals within the society (Finnveden 1999; Howard and Kneppers 2011). Those targets vary between countries since they are foremost politically based, and therefore results are dependent on the context. The monetary approach, e.g. the environmental priority strategies in product design method (Steen 1999), can be expressed through the willingness-to-pay principle, where a monetary value is assigned to the potential damage to environmental goods and services (Finnveden 1999). Since ways of calculating these real market prices are not available, the willingness-to-pay of different stakeholders is the basis for the weighting (Howard and Kneppers 2011).

An alternative approach to subjectively weighted single-score results could be the consistent presentation of un-weighted results in addition to weighted scores. Moreover, different weighting methods should be applied in case studies in order to analyze sensitivity of value choices taken.

3.3.4 Macroeconomic Scale-Up

Macroeconomic scale-up describes the expansion from a product or company perspective towards a country or regional perspective, including dynamic effects of production structures (e.g. production of one lithium battery vs. one million lithium batteries). Macroeconomic scale-up approaches aim to link environmental burdens on the micro level (e.g. product, process data) with information on the macro level (e.g. national, regional or sector data) (Reimann et al. 2010).

Process-based LCA is traditionally used to evaluate the environmental impacts of a specific process or product, but hardly addressing macroeconomic effects. As an example, production structures (mass production vs. niche production) and industrial dynamics (e.g. technological progress) are typically not considered within process-based LCA so far (Risku-Norja and Mäenpää 2007; Zhai and Williams 2010). The analysis of environmental impacts by means of process-based LCA may neglect significant parts of the overall environmental burden by ignoring macroeconomic infrastructures (e.g. road construction, machinery) (Frischknecht et al. 2007). The implementation of a macro level perspective can also be relevant for biofuels, bioplastics and other agricultural products, as the total amount of products might be limited due to limited availability of natural resources (Bringezu et al. 2009; Piemonte and Gironi 2012).

The two methods mainly used when considering macroeconomic perspective in connection with LCA are the hybrid LCA and the Basket-of-Products approach. Hybrid LCA combines a bottom-up approach based on facility-micro-level data (process-based LCA) with a top-down economic input-output (EIO) model to account for unavailable indirect macro-level data (Deng et al. 2011; Rugani et al. 2012; Strømman and Hertwich 2004; Suh et al. 2004; Suh and Nakamura 2007). However, the application of hybrid LCA is not obvious, as miscalculations may lead to double counting or leaked emissions (UNEP/SETAC 2011a). Even though the model aims at creating a more complete system by using additional inventory data and combining the macro and micro level data (Jeswani et al. 2010), hybrid LCA fails in including macroeconomic market dynamics. The Basket-of-Products method has been developed by the European Commission Joint Research Centre (JRC) for analyzing environmental burdens associated with a representative product of one sector. The approach matches macro level data on private consumption per capita with LCI micro level data for single products consumed. It reflects the environmental impact and the resource use associated with the final consumption of an average citizen in the EU-27 over the entire life cycle of goods and services. Direct environmental effects of consumption behavior in the corresponding sectors are displayed, based on apparent domestic final consumption and demand (EC 2012a). However, so far this approach has only been tested within one pilot study for five main sectors (nutrition, shelter, consumer goods, mobility, and services) in Germany and does not yet deliver a comprehensive solution to address macro-scale-up from a global perspective. The Basket-of-Products approach can serve as a basis to include the macro perspective into process LCA, even though industrial dynamics are not included. In addition, this approach could be complemented by macro level tools like EIO (environmental input-output)/hybrid LCA, MFA (material flow analysis) or related indicators like EMC (environmentally weighted material consumption) (Guinée et al. 2006; Risku-Norja and Mäenpää 2007; van der Voet et al. 2009).

3.3.5 Modeling Approach: Consequential LCA

LCA studies can be categorized into two general types: attributional (ALCA) and consequential (CLCA). The large majority of the LCAs today use the attributional modeling focusing on the total emissions during the life cycle of a product. The effects of changes within a life cycle are not considered. CLCA seeks to assess these changes (either positive or negative) in total emissions which result from a marginal change in the level of output of a product (EC-JRC 2010c) to inform about the consequence of decisions (e.g. the effects of an increase in milk production on soybean production) (Thomassen et al. 2008). CLCA tries to model the consequence of one additional unit of output rather than the average consequences of a product.

However, CLCA also adds uncertainty on top of existing gaps in LCA, as the effects of changes assessed in CLCA depend on economic mechanisms, complex economic models and market predictions representing relationships between demand for inputs, market effects, etc. (Ekvall 2002). Furthermore, due to the limited application in case studies, practitioners lack a common understanding of how to model CLCAs, and diverse understandings and procedures exist. Awareness and consideration of these current shortcomings and challenges are relevant as increasing attention is paid to consequential modeling by LCA practitioners.

Some case studies aim at doing CLCAs and assessing the effects of changes in a system (e.g. Thomassen et al. 2008; Schmidt and Weidema 2008; Dalgaard et al. 2008; Reinhard and Zah 2009; Ekvall and Andrae 2006). However, existing case studies and used models rarely provide high levels of accuracy, completeness or precision. Moreover, case studies just reflect a few effects, as CLCAs cannot describe the full consequences of a change (Ekvall 2002). Additionally, significant double counting of emissions can occur (Brander et al. 2009) as the scope of different CLCAs case studies may overlap, and the same emissions may be accounted for in multiple CLCAs.

Consultation of other methods can improve CLCA models: dynamic optimization models can improve knowledge on marginal effects; partial equilibrium models can improve knowledge of what product flows are affected by a change; general equilibrium models can give insight on rebound effects (Ekvall 2002). However, current application of CLCA can lead to suboptimal systems as the use of marginal data can lead to wrong incentives. As the future is inherently uncertain and significant limits exist to comprehensively describe future consequences of a change (Ekvall 2002), it is proposed to assess changes by means of a baseline using ALCA and different scenarios instead of using CLCA until more robust and consistent methods and case study experience are available.

3.3.6 Rebound Effects

Rebound effects refer to the change of environmental impacts when the implementation of an improvement option liberates or binds a scarce production or consumption factor (e.g. money, time, space) which can offset the effect of the measures (Schmidt and Weidema 2008; Weidema et al. 2008; Hertwich 2005; Schettkat 2009; Spielmann et al. 2008). Rebound effects can occur due to changes in production and consumption, e.g. when energy savings resulting from a more energy efficient option are cancelled out by increasing overall energy demand (Khazzoom 1980) or when the time saved by using faster and more efficient trains is used to travel further (Spielmann et al. 2008).

Although rebound effects are already considered in some LCA case studies, it is far from being common practice. Assessing and quantifying the changes in production and consumption can become very complex and challenging as, for example, a regional perspective (Spielmann et al. 2008) or personal behavior have to be included. Ignoring rebound effects leads to either under- or over-estimations of the

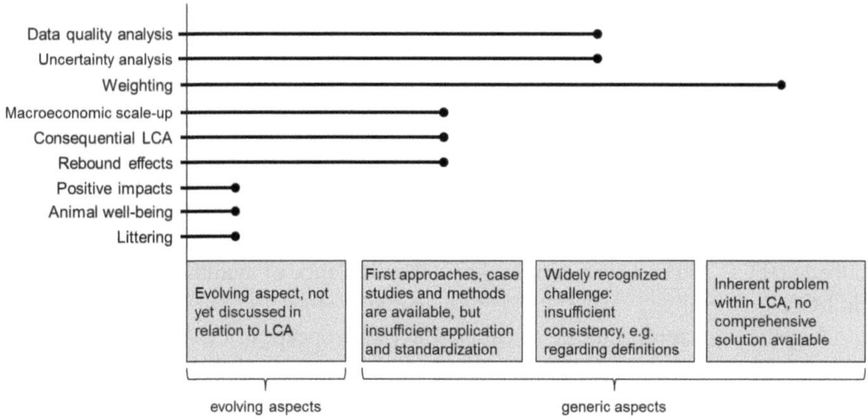

Fig. 7.5 Characteristics of 'generic' and 'evolving' challenges for LCA

impacts of products (Spielmann et al. 2008). Hence, for a realistic assessment of the environmental impacts related to a decision, rebound effects should ideally be included in LCA. This modeling of consequences of a decision is important for including a macro level perspective into LCA as well as to assess potential positive effects of products/processes.

Several approaches exist for the consideration of rebound effects within LCA. Finnveden et al. (2009) and Ibenholt (2002) proposed to use general equilibrium models, which can provide insight in rebound effects, and Ou et al. (2010) included scenario analysis in an attributional LCA case study. Furthermore, also the CLCA approach aims at considering rebound effects. However, several publications (e.g. Ekvall and Weidema 2004; Erikson et al. 2007; Frees 2008; Halleux et al. 2008) lead to the conclusion that the existing CLCA modeling (marginal or affected technology) does not address rebound effects properly.

As a good option for practitioners to capture the complexity of production and consumption, the use of (future) scenarios within attributional LCA is recommended by the European Commission (2010b) until robust and comprehensive macroeconomic scale up models are available that include rebound effects.

The main characteristics of gaps and challenges referring to 'generic aspects', described above, and 'evolving aspects', discussed in the following section, are summarized in Fig. 7.5.

3.4 *Evolving Aspects*

This section discusses aspects which are not yet been fully discussed in the context of LCA, such as the integration of 'positive impacts', 'animal well-being' and 'littering' in LCA.

3.4.1 Positive Impacts

Positive impacts refer to desirable or beneficial consequences or outcomes of product systems.

Current characterization models in LCA only consider negative environmental impacts, even though certain emissions can have a positive effect on the environment as well. One example is sulfur dioxide which reacts to sulfates, reducing global warming by increasing cloud generation (Brakkee et al. 2008) and by increased albedo (IPCC 2001). This example shows the relevance of including positive effects in LCA.

A systematic approach (the so called yin-yang concept of LCA) to identify positive (yin) and negative (yang) environmental impacts of substances and to include these impacts into the characterization models for each impact category, is proposed in Ackermann et al. (2009). The inclusion of potential positive effects of, e.g., substances was tested for the impact category human toxicity (Ackermann et al. 2010) based on data from a pharmaceutical database from Germany (ABDA 2012).

A broader set of potential positive impacts has to be identified and analyzed. It has to be explored in more detail how they can be taken into account in characterization models to enable comprehensive consideration in future LCAs.

3.4.2 Animal Well-Being

Existing definitions of animal well-being focus on how the animal is coping with its environment or stress, the fundamental behavioral needs that must be satisfied, or how animals should live according to their nature (Hewson 2003b). Many aspects of animal welfare are still unknown, and no broadly accepted definition of animal well-being exists (Hewson 2003b; BMELV 2012). When addressing animal well-being, two different types of ecosystems have to be distinguished: the ecosystem that is naturally available (e.g. natural forests), and the ecosystems created by humans (e.g. animal farm, managed forests). Regarding natural ecosystems, animal well-being could address the impacts that products and processes have on animals in the ecosystem, beyond existing impact categories like ecotoxicity or land use. With regard to 'agricultural ecosystems', animals can be considered as products, and animal well-being would refer to impacts on the products as such, in contrast to other categories that 'only' consider impacts of products on the environment.

So far, aspects influencing animal well-being are not included in the inventory or impact assessment within LCA. Generally, it could be considered as being outside the scope of LCA and rather being part of other methods, such as EIA or—with regard to the increasing consumer's interest in adequate animal husbandry—even SLCA. However, it could be interesting and beneficial to also include animal well-being into the (environmental) LCA discussion. LCA studies comparing conventional with organic animal farming mostly evaluate conventional systems as more efficient and environmentally friendly (per unit of output), but differences in animal treatment (between organic and conventional animal farming) are not included within the assessment (Hospido et al. 2003; Thomassen et al. 2009).

Some studies focusing on the development of indicators and frameworks to assess animal well-being related to farming are already available (Hewson 2003a; Deimel et al. 2010; de Vries et al. 2011; Hofmann et al. 2000; Häusler and Scherer-Lorenzen 2002). However, integration or combination with LCA was not yet proposed. For an inclusion of animal well-being in LCA, indicator development should consider the two aforementioned ecosystem classifications. Due to an increasing number of agricultural LCAs, indicator development for animal farming could be a priority. Indicators available in existing studies (e.g. to assess space or naturalness of feed) can be used as starting point for indicator development regarding animal well-being in LCA. For instance, weighting factors based on the distance-to-target principle could be developed. By relating the current situation of, for example, space per animal or open-air access to target values derived from organic farming standards, exceedance of animal well-being thresholds could be quantified. To establish animal well-being as a new impact category, further research is needed to clearly define the impact pathways, to enhance existing indicators, to better understand feasibility of these indicators, and to possibly also broaden the scope to address both ecosystems related to animal well-being.

3.4.3 Littering

Littering describes the disposal of products directly into the environment without any waste treatment. Consequences of littering comprise, for example, leaching of chemicals from electronic devices and batteries, cause of fires by glass or cigarettes, killing of animals and fish when eating plastics (Wong et al. 1974), and aesthetic disturbance of the landscape. Littering is caused by human behavior, e.g. due to the lack of proper education, as well as the absence of adequate regulations in many countries (Gamarra and Salhofer 2007).

As LCA has historically considered the intended way of disposal only; littering has not been addressed on the LCI or on the LCIA level. Hence, LCA neglects the fact that some products are more likely to be littered (e.g. beverage cartons) than others (e.g. deposit bottles) and that consequences of littering might differ between products (e.g. glass and plastic bottles).

In order to address this shortcoming, littering could be assessed by determining the percentage of a product being disposed of improperly. The consequences of littering could be evaluated by means of existing or new impact categories. For instance, leaching of chemicals from a battery could be assessed by means of human- and ecotoxicity impact categories. Forest fires caused by glass or cigarette littering could be assessed by land use and global warming characterization models. The killing of animals due to swallowing of plastic waste could be evaluated by means of a new impact category 'direct non-intended killing of animals' as proposed in Sect. 3.2.2.4.

4 Conclusion

This chapter summarizes the content, relevance, state of the art literature, and potential solutions of 34 gaps and challenges of LCA identified. They encompass 'inventory aspects' (see Sect. 3.1), 'impact assessment aspects' (see Sect. 3.2), 'generic aspects' (see Sect. 3.3), and 'evolving aspects' (see Sect. 3.4). Despite the large number and the broad range of these challenges, the overall scientific robustness achieved in LCA needs to be acknowledged first.

LCA can assist in identifying opportunities to improve the environmental performance of products at various points of their life cycle. Moreover, it enables fact-based decision support for industry, government or non-government organizations (e.g. for the purpose of strategic planning, priority setting, product or process design, environmental hot spot analysis, and optimization). Despite the large number of gaps and challenges identified, LCA is still the "…best framework for assessing the potential environmental impacts of products currently available" (EU 2003).

However, the methodological gaps and challenges can have a significant influence on the results of LCA studies, even though not every individual case study suffers from all 34 gaps. Some of the challenges discussed above, e.g. 'nanomaterials', 'animal well-being' or 'littering', are only relevant for particular applications or products. As a consequence, they do not represent severe issues for most case studies. However, if LCAs for plastic packaging or livestock production are performed, the absence of a proper coverage of 'littering' or 'animal well-being' excludes potentially significant issues from the assessment.

A number of challenges, e.g. 'allocation', 'functional unit' or 'uncertainty analysis', is inherent to the LCA method as such. While many of the challenges identified above can be addressed by future scientific work and progress, these fundamental challenges may inherently require value choices. These choices can be scientifically informed, but they remain value choices. On a scientific level, it can only be checked as to whether these value choices are made consistently throughout a study and particularly between alternatives.

The conclusion of this study for the scientific community is rather self-explanatory. The challenges described can be used as a research agenda for LCA. Most gap descriptions include some specific proposals for further research. They are intended to motivate the scientific LCA community for tackling these challenges and to inspire the development of scientifically robust solutions. A recurrent topic for many challenges identified is the need for additional, robust and relevant data. This is a task for stakeholders, not only for science.

Until these solutions are developed, users and decision makers face a slightly more complex situation, as the relevance of the gaps depends on the products studied and the intended application of the LCA. As a consequence, each case study should include a check, if and how the identified gaps influence or even limit the conclusions of a particular LCA study. For this purpose, Fig. 7.1 could be used as a kind of checklist for case studies. If the gaps summarized in Fig. 7.1

are screened against a specific case study context, the scope definition of this study is better informed. For each case study, some gaps will turn out to be irrelevant, some will be of minor relevance and some might be significant. The potentially significant issues should be either documented as limitation of the scope of the study, or they should be tackled with complementary tools like, e.g., risk assessment, environmental impact assessment, material flow analysis or input-output analyses.

Decision-makers in both private and public organizations need to appreciate the benefits of LCA. However, a robust, sustainable and credible use of LCA requires a proper consideration of its gaps and limitations. LCAs should be seen as one relevant element of environmentally motivated decision making, but as ISO 14044 (2006) puts it: "An LCIA shall not provide the sole basis of...overall environmental superiority or equivalence, as additional information will be necessary to overcome some of the inherent limitations in the LCIA."

Acknowledgements We appreciate the support of the German Research Foundation DFG for funding some of the research presented here as part of the Collaborative Research Center 'Sustainable Manufacturing' (SFB 1026).

References

ABDA (2012) ABDATA Pharma-Daten-Service. http://www.dimdi.de/static/de/amg/abda/index.htm. Accessed 29 April 2013

Ackermann R, Bergmann T, Finkbeiner M (2009) Yin-/Yang-concept of environmental sustainability assessment. In: Shunmugam MS, Babu NR (eds) 7th global Conference on sustainable manufacturing, Chennai

Ackermann R, Bergmann T, Jung K, Finkbeiner M (2010) Yin-/Yang-concept for life cycle impact assessment. In: Life Cycle Assessment X, Portland, Oregon

Althaus H-J, de Haan P, Scholz RW (2009) Traffic noise in LCA, Part 1: state-of-science and requirement profile for consistent context-sensitive integration of traffic noise in LCA. Int J Life Cycle Assess 14(6):560–570

Ambrose EE, Solarin BB, Isebor CE, Williams AB (2005) Assessment of fish by-catch species from coastal artisanal shrimp beam trawl fisheries in Nigeria. Fish Res 71(1):125–132. doi:10.1016/j.fishres.2004.07.005

Amores-Barrero MJ, Verones F, Raptis C, Juraske R, Pfister S, Stoessel F, Antón A, Castells F, Hellweg S (2013) Biodiversity impacts from salinity increase in a coastal wetland. Environ Sci Technol 47(12):6384–6392. doi:10.1021/es3045423

Anderson-Teixeira KJ, Snyder PK, De Lucia EH (2011) Do biofuels life cycle analyses accurately quantify the climate impacts of biofuels-related land use change? Univ Ill Law Rev 2:589–622

ANEC (2012) ANEC position: environmental assessment goes astray. A critique of environmental footprint methodology and its ingredients. ANEC-ENV-2012-G-008final. ANEC, the European association for the co-ordination of consumer representation in standardisation. http://www.anec.eu/attachments/ANEC-ENV-2012-G-008final%20%283%29.pdf. Accessed 20 Mar 2012

Ardente F, Beccali G, Cellura M, Lo Brano V (2005) Life cycle assessment of a solar thermal collector. Renew Energ 30(7):1031–1054. doi:http://dx.doi.org/10.1016/j.renene.2004.09.009

Ardente F, Beccali M, Cellura M, Lo Brano V (2008) Energy performances and life cycle assessment of an Italian wind farm. Renew Sust Eng Rev 12(1):200–217. doi:http://dx.doi.org/10.1016/j.rser.2006.05.013

Arnett EB, Strickland MD, Morrison ML (2007) Renewable energy resources and wildlife: impacts and opportunities. Transactions of the 72nd North American Wildlife and Natural Resources Conference

Arvesen A, Hertwich EG (2012) Assessing the life cycle environmental impacts of wind power: a review of present knowledge and research needs. Renew Sust Eng Rev 16(8):5994–6006. doi:http://dx.doi.org/10.1016/j.rser.2012.06.023

Azapagic A, Clift R (1999) Allocation of environmental burdens in multiple-function systems. J Clean Prod 7:101–119

Baker J, Lepech M (2009) Treatment of uncertainties in life cyle assessment. Stanford University, Stanford

Balmford A, Bennun L, ten Brink B, Cooper D, Cote IM, Crane P, Dobson A, Dudley N, Dutton I, Green RE, Gregory RD, Harrison J, Kennedy ET, Kremen C, Leader-Williams N, Lovejoy TE, Mace G, May R, Mayaux P, Morling P, Phillips J, Redford K, Ricketts TH, Rodriguez JP, Sanjayan M, Schei PJ, van Jaarsveld AS, Walther BA (2005) The convention on biological diversity's 2010 target. Science 307(5707):212–213. doi:10.1126/science.1106281

Banks RC (1979) Human related mortality of birds in the United States. Washington, DC

Bayart JB, Bulle C, Koehler A, Margni M, Pfister S, Vince F, Deschenes L (2010) A framework for assessing off-stream freshwater use in LCA. Int J Life Cycle Assess 15(5):439–453

Becker H, Herzberg F, Schulte A, Kolossa-Gehring M (2010a) The carcinogenic potential of nanomaterials, their release from products and options for regulating them. Int J Hyg Environ Health 214(3):231–238

Becker M, Edwards S, Massey RI (2010b) Toxic chemicals in toys and children's products: limitations of current responses and recommendations for government and industry. Environ Sci Technol 44(21):7986–7991

Benetto E, Dujet C (2003) Uncertainty analysis and MCDA: a case study from the life cycle assessment (LCA) practice. In: 57th Meeting of the European working group on multicriteria decision aiding, Viterbo, 27–29 March

Berger M, Finkbeiner M (2010) Water footprinting: how to address water use in life cycle assessment? Sustainability 2:919–944. doi:10.3390/su2040919

Berger M, Warsen J, Krinke S, Bach V, Finkbeiner M (2012) Water footprint of European cars: potential impacts of water consumption along automobile life cycles. Environ Sci Technol 46(7):4091–4099

Bindraban P, van der Velde M, Ye L, van den Berg M, Materechera S, Kiba D, Tamene L, Vala RK, Jongschaap R, Hoogmoed M, Hoogmoed W, van Beek C, van Lynden G (2012) Assessing the impact of soil degradation on food production. Current Opinion in Environmental Sustainability 4(5):478–488

Birnbaum LS, Jung P (2011) From endocrine disruptors to nanomaterials: advancing our understanding of environmental health to protect public health. Health Aff 30(5):814–822. doi:10.1377/hlthaff.2010.1225

Blaisdell R (2007) Health effects of odors. Exposure modeling section. Office of Enviromental Health Hazard Assessment. https://www.aqmd.gov/tao/conferencesworkshops/OdorForum/BlaisdellSlides.pdf. Accessed 25 May 2013

BMELV (2012) Charta für Landwirtschaft und Verbraucher. Bundesministerin für Ernährung, Landwirtschaft und Verbraucherschutz. doi:http://dx.doi.org/10.1016/S0140-6736(02)11274-8

Boulay A-M, Bouchard C, Bulle C, Deschenes L, Margni M (2011) Categorizing water for LCA inventory. Int J Life Cycle Assess 16(7):639–651

BPA (2013) Integrated fish and wildlife program. http://efw.bpa.gov/IntegratedFWP/. Accessed 6 Feb 2013

Brakkee K, Huijbregts M, Eickhout B, Hendriks A (2008) Characterisation actors for greenhouse gases at a midpoint level including indirect effects based on calculations with IMAGE model. Int J Life Cycle Assess 13(3):191–201

Brander M, Tipper R, Hutchison C, Davis G (2009) Consequential and attributional approaches to LCA: a guide to policy makers with specific reference to greenhouse gas LCA of biofuels Technical Paper TP-090403-A. Ecometrica

Bringezu S, Schütz H, O'Brien M, Kauppi L, Howarth RW, McNeely J (2009) Towards sustainable production and use of resources: assessing biofuels. United Nations Environment Programme (UNEP)
Brunekreef B, Holgate ST (2002) Air pollution and health. The Lancet 360(9341):1233–1242. doi:http://dx.doi.org/10.1016/S0140-6736(02)11274-8
Burger AE (1993) Estimating the mortality of seabirds following oil-spills-effects of spill volume. Mar Pollut Bull 26(3):140–143. doi:10.1016/0025-326x(93)90123–2
Burkhardt JJ, Heath G, Cohen E (2012) Life cycle greenhouse gas emissions of trough and tower concentrating solar power electricity generation. J Ind Ecol 16:93–109. doi:10.1111/j.1530-9290.2012.00474.x
Canada Mortgage and Housing Corporation (2004) Sensitivity and uncertainity. vol Annex 31. Energy-related environmental impact of buildings. International Energy Ageny, Energy conservaton in buildings and bommunity systems programme
Cherubini F, Bird ND, Cowie A, Jungmeier G, Schlamadinger B, Woess-Gallasch S (2009) Energy-and greenhouse gas-based LCA of biofuel and bioenergy systems: key issues, ranges and recommendations. Res Con Rec 53(8):434–447. doi:http://dx.doi.org/10.1016/j.resconrec.2009.03.013
Ciroth A, Fleischer G, Steinbach J (2004) Uncertainty calculation in life cycle assessments—a combined model of simulation and approximation. Int J Life Cycle Assess 9(4):216–226
Clark C, Martin R, van Kempen E, Alfred T, Head J, Davies HW, Haines MM, Barrio IL, Matheson M, Stansfeld SA (2006) Exposure-effect relations between aircraft and road traffic noise exposure at school and reading comprehension—The RANCH project. Am J Epidemiol 163(1):27–37
Cleuvers M (2003) Aquatic ecotoxicity of pharmaceuticals including the assessment of combination effects. Toxicol Lett 142(3): 185–194. doi:10.1016/s0378-4274(03)00068–7
Cole SG (2011) Wind power compensation is not for the birds: an opinion from an environmental economist. Restoration Ecol 19(2):147–153
Cooper JS (2003) Specifying functional units and reference flows for comparable alternatives. Int J Life Cycle Assess 8(6):337–349
Cornman D (2003) Effects of noise on wildlife. Nature Sounds Society. http://www.naturesounds.org/conservENW.html. Accessed 21 May 2013
Cowell SJ, Clift R (2000) A methodology for assessing soil quantity and quality in life cycle assessment. J Clean Prod 8(4):321–331
Cucurachi S, Heijungs R, Ohlau K (2012) Towards a general framework for including noise impacts in LCA. Int J Life Cycle Assess 17(4):471–487. doi:10.1007/s11367-011-0377-4
Curran MA (2007) Co-product and input allocation approaches for creating life cycle inventory data—a literature review. Int J Life Cycle Assess 12(1):65–78
Curran MA, Mann M, Norris GA (2001) Report on the international workshop on electricity data for life cycle inventories Available via EPA http://nepis.epa.gov/Adobe/PDF/P1001NRO.pdf. Accessed 28 May 2013
Curran M, de Baan L, de Schryver AM, van Zelm R, Hellweg S, Koellner T, Sonnemann G, Huijbregts MAJ (2011) Toward meaningful end points of biodiversity in life cycle assessment. Environ Sci Technol 45(1):70–79
Dalgaard R, Schmidt J, Halberg N, Christensen P, Thrane M, Pengue WA (2008) LCA of soybean meal. Int J Life Cycle Assess 13(3):240–254
de Vries M, Bokkers E, Dijkstra T, von Schaik G, de Boer J (2011) Invited review: Associations between variables of routine herd data and dairy cow cattle wellfare indicators. J Dairy Sci 94:3213–3228
Deimel I, Franz A, Frentrup M, Meyer M von, Spiller A, Theuvsen L, Dettmer J, Gauly M, Salzborn C, Schrader L, Weghe H van den (2010) Perspektiven für ein Europäisches Tierschutzlabel. Available via BLE http://download.ble.de/08HS010.pdf. Accessed 29 May 2013
Delfino RJ, Sioutas C, Malik S (2005) Potential role of ultrafine particles in associations between airborne particle mass and cardiovascular health. Environ Health Perspectives 113(8):934–946
Deng L, Babbitt CW, Williams ED (2011) Economic-balance hybrid LCA extended with uncertainty analysis: case study of a laptop computer. J Clean Prod 19(11):1198–1206. doi:10.1016/j.jclepro.2011.03.004

Diamanti-Kandarakis E, Bourguignon J-P, Giudice LC, Hauser R GSP, Soto AM et al (2009) Endocrine-disrupting chemicals: an Endocrine Society scientific statement. Endocrine Rev 30(4):293–342

Dodić SN, Popov SD, Dodić JM, Ranković JA, Zavargo ZZ (2010) Biomass energy in Vojvodina: market conditions, environment and food security. Renew Sust Energ Rev 14(2): 862–867. doi:http://dx.doi.org/10.1016/j.rser.2009.10.019

Dovì VG, Friedler F, Huisingh D, Klemeš JJ (2009) Cleaner energy for sustainable future. J Clean Prod 17(10):889–895. doi:http://dx.doi.org/10.1016/j.jclepro.2009.02.001

EC (2010) Critical raw materials for the EU. Report of the Ad-hoc Working Group on defining critical raw materials. Raw Materials Supply Group, European Commission

EC (2012a) Decoupling indicators: basket-of-products indicators, waste management indicators, framework, methodology, data basis and updating procedures

EC (2012b) Endocrine disruptors website. European Commission. http://ec.europa.eu/environment/endocrine/definitions/endodis_en.htm. Accessed 21 May 2013

EC (2012c) Nanomaterials. http://ec.europa.eu/environment/chemicals/nanotech/index.htm. Accessed 11 February 2013

EC-JRC (2010a) International reference life cycle data system (ILCD) handbook—framework and requirements for LCIA models and indicators. EUR 24586 EN. European Commission – Joint Research Centre – Institute for Environment and Sustainability. Publications Office of the European Union

EC-JRC (2010b) International reference life cycle data system (ILCD) handbook—General guide for life cycle assessment—Detailed guidance. EUR 24709 EN. European Commission—Joint Research Centre—Institute for Environment and Sustainability

EC-JRC (2010c) International reference life cycle data system (ILCD) handbook—Specific guide for life cycle inventory data sets. EUR 24709 EN. European Commission—Joint Research Centre—Institute for Environment and Sustainability

EC-JRC (2011) International reference life cycle data system (ILCD) handbook—Recommendations for life cycle impact assessment in the European context. EUR 24571 EN, vol EUR 24571 EN., 1 edn. European Commission—Joint Research Centre—Institute for Environment and Sustainability. Publications Office of the European Union, Luxemburg

Ecoinvent (2013) Ecoinvent 2.2 and 3.0 Swiss centre for Life Cycle Inventories. Available via http://www.ecoinvent.ch/

EEA (2012) The impacts of endocrine disrupters on wildlife, people and their environments. European Environmental Agency (EEA) Technical Report. doi:10.2800/41462

EFSA (2013) EFSA supports European Commission in defining scientific criteria for endocrine disruptors. Press release 20 March. European Food Safety Authority http://www.efsa.europa.eu/en/press/news/130320.htm. Accessed 21 May 2013

Ekvall T (2002) Limitations of consequential LCA. http://www.lcacenter.org/lca-lcm/pdf/Consequential-LCA.pdf. Accessed 20 Mar 2013

Ekvall T, Andrae ASG (2006) Attributional and consequential environmental assessment of the shift to lead-free solders. Int J Life Cycle Assess 11(5):344–353

Ekvall T, Finnveden G (2001) Allocation in ISO 14041—a critical review. J Clean Prod 9:197–208

Ekvall T, Finnveden G (2009) Allocation in ISO 14041—a critical review. J Clean Prod 9:197–208

Ekvall T, Tillmann AM (1997) Open-loop recycling: criteria for allocation procedures. Int J Life Cycle Assess 2(3):155–162

Ekvall T, Weidema B (2004) System boundaries and input data in consequential life cycle iInventory analysis. Int J Life Cycle Assess 9(3):161–171

EPA (1993) Life cycle assessment: Inventory guidelines and principles. Cincinnati

EPA (2012) Environmental databases. http://www.epa.gov/opp00001/science/efed_databasesdescription.htm#ecotoxicity. Accessed 25 Feb 2013

Erdmann L, Graedel TE (2011) Criticality of non-fuel minerals: a review of major approaches and analyses. Environ Sci Technol 45(18):7620–7630

Erickson WP, Johnson GD, Young DP (2005) A summary and comparison of bird mortality from anthropogenic causes with an emphasis on collisions. USDA Forest Service 1029–1042

Erikson O, Finnveden G, Ekvall T, Bjorklund A (2007) Life cycle assessment of fuels for district heating: a comparison of waste incineration, biomass- and natural gas combustion. Eng Pol 35(17):1346–1362

Ernstoff A (2012) Direct consumer exposure during use of personal care products, plasticizers and flooring materials. Paper presented at the 9th International Conference on EcoBalance, Yokohama, Japan

EU (2003) European Union (EU): Communication from the Commission to the Council and the European Parliament of 18 June 2003– integrated product policy—building on environmental life-cycle thinking. COM (2003) 302 final. Brussels, Belgium

EU (2009) Nanomaterials. European Union (EU), DG Health and Consumer Protection, Public Health. http://ec.europa.eu/health/scientific_committees/opinions_layman/nanomaterials/en/index.htm#4. Accessed 21 May 2013

Feitz AJ, Lundie S (2002) Soil salinisation: a local life cycle assessment impact category. Int J Life Cycle Assess 7(4):244–249

Ferrer M, de Lucas M, Janss GFE, Casado E, Munoz AR, Bechard MJ, Calabuig CP (2012) Weak relationship between risk assessment studies and recorded mortality in wind farms. J Appl Ecol 49(1):38–46. doi:10.1111/j.1365-2664.2011.02054.x

Finkbeiner M (2009) Carbon footprinting—opportunities and threats. Int J Life Cycle Assess 14:91–94

Finkbeiner M (2013) Indirekte Landnutzungsänderungen in Ökobilanzen—wissenschaftliche Belastbarkeit und Übereinstimmung mit internationalen Standards. Technische Universität Berlin, Lehrstuhl am Fachgebiet Sustainable Engineering am Institut für Technischen Umweltschutz

Finkbeiner M, Schau EM, Lehmann A, Traverso M (2010) Towards life cycle sustainability assessment Sustainability 2: 3309–3322

Finkbeiner M, Neugebauer S, Berger M (2012) Carbon footprint of recycle biogenic products: the challenge of modelling CO_2 removal credits. Int J Sustain Eng 1(6):1–8

Finnveden G (1999) A critical review of operational valuation/weighting methods for life cycle assessment. Available via AFN http://naturvardsverket.se/Documents/publikationer/afr-r-253-se.pdf. Accessed 28 May 2013

Finnveden G, Hauschild MZ, Ekvall T, Guinée J, Heijungs R, Hellweg S, Koehler A, Pennington D, Suh S (2009) Recent developments in life cycle assessment. J Environ Management 91(1):1–21

Fleet D, Franeker JV, Dagevos J, Hougee M (2009) Quality status report—marine Litter. Wilhelmshaven

Flysjo A, Cederberg C, Henriksson M, Ledgard S (2012) The interaction between milk and beef production and emissions from land use change—critical considerations in life cycle assessment and carbon footprint studies of milk. J Clean Prod 28:134–142. doi:10.1016/j.jclepro.2011.11.046

Franco V, Garraín D, Vidal R (2010) Methodological proposals for improved assessments of the impact of traffic noise upon human health. Int J Life Cycle Assess 15(8):869–882

Frees N (2008) Crediting aluminium recycling in LCA by demand of by disposal. Int J Life Cycle Assess 13(3):212–218

Frischknecht R (2010) LCI modelling approaches applied on recycling of materials in view of environmental sustainability, risk perception and eco-efficiency. Int J Life Cycle Assess 15(7):666–671

Frischknecht R, Althaus H, Bauer C, Doka G, Heck T, Jungbluth N, Kellenberger D (2007) The environmental relevance of capital goods in life cycle assessments of products and services. Int J Life Cycle Assess 12(1):1–11. doi: http://dx.doi.org/10.1065/lca2007.02.309 [PDF via treeze.ch]

Frischknecht R, Steiner R, Jungbluth N AB (2009) The ecological scarcity method—Eco-factors 2006. A method for impact assessment in LCA. Environmental Studies no 0906. Federal Office for the Environment (FOEN), Bern

Gamarra P, Salhofer S (2007) A comparison of waste management in Peru and some Latin-American countries: an overview of major problems, characteristiccs and real needs in the region. In: Diaz LF, Eggerth LL, Savage GM (eds) Management of solid wastes in developing countries. IWWG Task group on waste management in developing countries. CISA Publisher, Padova, pp 71–81

Garrigues E, Corson MS, Angers DA, van der Werf HMG, Walter C (2012) Soil quality in life cycle assessment: towards development of an indicator. Ecol Indic 18:434–442. doi:10.1016/j.ecolind.2011.12.014

Garrigues E, Corson M, Angers D, Werf HG, Walter C (2013) Development of a soil compaction indicator in life cycle assessment. Int J Life Cycle Assess 18:1316–1324

Garrity SD, Levings SC (1993) Effects of an oil-spill on some organisms living on mangrove (rhizophora-mangle L) roots in low wave-energy habitats in Carribean Panama. Mar Environ Res 35(3): 251–271. doi:10.1016/0141-1136(93)90097-j

Gavankar S, Suh S, Keller AF (2012) Life cycle assessment at nanoscale: review and recommendations. Int J Life Cycle Assess 17(3):295–303. doi:10.1007/s11367-011-0368-5

Geyer R, Lindner JP, Stoms DM, Davis FW, Wittstock B (2010) Coupling GIS and LCA for biodiversity assessments of land use: part 2: impact assessment. Int J Life Cycle Assess 15(7):692–703

GHG (2011) Greenhouse gas protocol—Life cycle accounting and reporting standard. World Business Council for Sustainable Development, World Resources Institute

Goedkoop MJ, Spriensma R (2001) The eco-indicator 99– A damage oriented method for life cycle impact assessment. vol 3. PRé Consultants B.V., Amersfoort, The Netherlands

Goedkoop MJ, Müller-Wenk R, Mettier T, Hungerbühler K, Braunschweig A, Klaus T (2000) Eco-indicator 99– Eine schadensorientierte Bewertungsmethode. Paper presented at the Nachbereitung zum 12 Diskussionsforum Ökobilanzen vom 30 Juni 2000 an der ETH Zürich

Goedkoop MJ, Heijungs R, Huijbregts M, de Schryver A, Struijs J, van Zelm R (2009) ReCiPe 2008– A life cycle impact assessment method which comprises harmonised category indicators at the midpoint and the endpoint level. Den Haag

Gontier M, Balfors B, Mörtberg U (2006) Biodiversity in environmental assessment Amersfort Ccurrent practice and tools for prediction. Environ Impact Assess Rev 26(3):268–286

Gottschalk F, Sonderer T, Scholz RW, Nowack B (2009) Modeled environmental concentrations of engineered nanamaterials for different regions. Environ Sci Technol 43:9126–9222

Graedel T, Barr R, Chandler C, Chase T, Choi J, Christofferson L, Friedlander E, Henly C, Jun C, Nassar N, Schechner D, Warren S, Yang M-y, Zhu C (2012) Methodology of metal criticality determination. Environ Sci Technol 46(2):1063–1070. doi: 10.1021/es203534z

Greco SL, Wilson AM, Spengler JD, Levy JI (2007) Spatial patterns of mobile source particulate matter emissions-to-exposure relationships across the United States. Atmos Environ 41(5):1011–1025

Griefahn B, Marks A, Robens S (2006) Noise emitted from road, rail and air traffic and their effects on sleep. J Sound Vibration 295(1–2):129–140

Grießhammer R, Hochfeld C (2009) Memorandum product carbon footprint. Öko-Institut e.V., UBA und BMU, Freiburg

Guinée JB, Gorree M, Heijungs R, Huppes G, Kleijn R, de Koning A, van Oers L, Wegener Sleeswijk A, Suh S, Udo deHHA, de Bruijn H, van Duin R, Huijbregts MAJ (2002) Handbook on life cycle assessment. Operational guide to the ISO standards. I: LCA in perspective. IIa: Guide. IIb: Operational annex. III: Scientific background, vol 7. Kluwer Academic Publishers, Dordrecht, The Netherlands

Guinée JB, Heijungs R, Huppes G (2004) Economic allocation: examples and derived decision tree. Int J Life Cycle Assess 9(1):23–33. doi:10.1007/bf02978533

Guinée JB, Heijungs R, Kleijn R, Voet EVD, Koning AD, Oers LV, Elshkaki A, Huele R, Huppes G, Suh S, Sleeswijk AW (2006) Human and ecological life cycle tools for the integrated assessment of systems (HELIAS). Int J Life Cycle Assess 11(1):19–28

Guney M, Zagury GJ (2011) Toxic chemicals in toys and children's products. Environ Sci Technol 45(9):3819–3819. doi:10.1021/es200810s

Halleux H, Lassaux S, Renzoni R, Germain A (2008) Comparative life cycle assessment of two biofuels: ethanol from sugar beet and rapeseed methyl ester. Int J Life Cycle Assess 13(3):184–190

Häusler A, Scherer-Lorenzen M (2002) Nachhaltige Forstwirtschaft in Deutschland im Spiegel des ganzheitlichen Ansatzes der Biodiversitätskonvention Bundesamt für Naturschutz, Bonn

HCES (2013) Particulate matter: little things can cause big problems. Hamilton County, Environmental services, airquality management division. http://www.hcdoes.org/airquality/monitoring/pm.htm#Environment. Accessed 20 Jan 2013

Heijungs R (2010) Uncertainty analysis in LCA—concepts, tools, and practice. Institute of Environmental Sciences (CML) Leiden University

Heijungs R, Guinée JB, Huppes G, Lankreijer RM, Udo deHHA, Wegener Sleeswijk A., Ansems AMM, Eggels PG, van Duin R, de Goede HP (1992) Environmental life cycle assessment of products. I: Guide. II: Backgrounds. Centre of environmental science (CML), Leiden University, Leiden

Heijungs R, Huijbregts M (2004) A review of approaches to treat uncertainty in LCA. In: Pahl-Wostl S, Schmidt AE, Rizzoli, Jakeman AJ (eds) Complexity and integrated resources management Transactions of the 2nd biennial meeting of the international environmental modelling and software society, vol 1. Osnabrück, p 1533

Henderson A, Hauschild M, van de Meent D, Huijbregts M, Larsen H, Margni M, McKone T, Payet J, Rosenbaum R, Jolliet O (2011) USEtox fate and ecotoxicity factors for comparative assessment of toxic emissions in life cycle analysis: sensitivity to key chemical properties. Int J Life Cycle Assess 16(8):701–709. doi:10.1007/s11367-011-0294-6

Henderson A, Ernstoff A, Chung S, Jolliet O (2012) Assessing direct impact of products on consumers in LCA. Paper presented at the LCA XII—Life cycle thinking ... life cycle living, Tacoma, 25–27 September

Henzen C (2008) The impact of land use on biodiversity in the framework of life cycle assessment. University of Basel, Basel

Hertwich EG (2005) Consumption and the rebound effect. J Ind Ecol 9(1–2):85–98

Hewson C (2003a) Can we assess welfare? Canad Vet J 44(9):749–753

Hewson C (2003b) What is animal welfare? Common definitions and their practical consequences. The Canad Vet J 44(6):496–499

Hischier R, Reichart I (2003) Multifunctional electronic media—Traditional media—The problem of an adequate functional unit—A case study of a printed newspaper, an internet newspaper and a TV broadcast. Int J Life Cycle Assess 8(4):201–208

Hofmann F, Kill J, Meder R, Plachter H, Volz K-R (2000) Waldnutzung in Deutschland Bestandsaufnahme, Handlungsbedarf und Maßnahmen zur Umsetzung des Leitbildes einer nachhaltigen Entwicklung. Metzler-Poeschel., Stuttgart

Hospido A, Moreira M, Feijoo G (2003) Simplified life cycle assessment of galician milk production. Int Dairy J 13:783–796

Howard N, Kneppers B (2011) Weighting for LCA based tools, methods and ecolabels—practical but contentious! http://www.conference.alcas.asn.au/2011/HowardKneppers.pdf. Accessed 29 Mar 2013

Hsu DD, O'Donoughue P, Fthenakis V, Heath GA, Kim HC, Sawyer P, Choi J-K, Turney DE (2012) Life cycle greenhouse gas emissions of crystalline silicon photovoltaic electricity generation. J Ind Ecol 16:122–135. doi:10.1111/j.1530-9290.2011.00439.x

Huijbregts M (1998) Application of uncertainty and variability in LCA. Int J Life Cycle Assess 3(5):273–280

Humbert S, Marshall JD, Shaked S, Spadaro JV, Nishioka Y, Preiss P, McKone TE, Horvath A, Jolliet O (2011) Intake fraction for particulate matter: recommendations for life cycle impact assessment. Environ Sci Technol 45(11):4808–4816

Hunkeler D, Lichtenvort K, Rebitzer G (eds) (2008) Environmental life cycle costing. SETAC, Pensacola, FL in collaboration with CRC Press, Boca Raton, FL

Huppes G, Oers Lv (2011) Background review of existing weighting approaches in life cycle impact assessment (LCIA). European Commission Joint Research Centre, Institute for Environment and Sustainability

Ibenholt K (2002) Materials flow analysis and economic modelling. In: Ayres RU, Ayres LW (eds) Handbook of industrial ecology. Edward Elgar, Cheltenham, pp 177–184

IPCC (2001) Third assessment report—climate change. Cambridge University Press

ISO 14040 (2006) Environmental management—Life cycle assessment—Principles and framework (ISO 14040:2006). International Organization for Standardization, Geneva, Switzerland

ISO 14044 (2006) Environmental management—Life cycle assessment—Requirements and guidelines (ISO 14044:2006). International Organization for Standardization, Geneva, Switzerland

ISO 31000 (2009) Risk management—Principles and guidelines (ISO 31000:2009). International Organization for Standardization, Geneva, Switzerland

ISO/TS 14067 (2013) Greenhouse gases—Carbon footprint of products—Requirements and guidelines for quantification and communication. International Organization for Standardization Geneva, Switzerland

ISO/TR 14049 (2012) Environmental management—Life cycle assessment—Illustrative examples on how to apply ISO 14044 to goal and scope definition and inventory analysis. International Organization for Standardization, Geneva, Switzerland

Jannick H Schmidt JR, Bo PW (2012) A model of indirect land use change. Proceedings of the 8th International Conference on Life Cycle Assessment in the Agri-Food Sector (LCA Food 2012), 1–4 October 2012, Saint Malo, France. INRA, Rennes, France, p 7

Jeffery S, Gardi C, Jones A (eds) (2010) European atlas of soil biodiversity. EUR 24375 EN European Commission—Joint Research Centre, Publication Office of the European Union, Luxemburg

Jeswani HK, Azapagic A, Schepelmann P, Ritthoff M (2010) Options for broadening and deepening the LCA approaches. J Clean Prod 18(2):120–127. doi:10.1016/j.jclepro.2009.09.023

Jolliet O, Margni M, Charles R, Humbert S, Payet J, Rebitzer G, Rosenbaum R (2003) IMPACT 2002 + : A new life cycle impact assessment methodology. Int J Life Cycle Assess 8(6):324–330. doi:10.1007/bf02978505

Jolliet O, Müller-Wenk R, Bare J, Brent A, Goedkoop M, Heijungs R, Itsubo N, Pena C, Pennington D, Potting J, Rebitzer G, Stewart M, Udo deHHA, Weidema B (2004) The LCIA midpoint-damage framework of the UNEP/SETAC life cycle initiative. Int J Life Cycle Assess 9(6):394–404

Jolliet O, Hong J, Rosenbaum R, Dettling J (2009) Analytical uncertainty propagation in life cycle inventory and impact assessment: high-efficiency versus conventional hand dryers. In: LCA IX—Toward the global life cycle economy, Boston, 29 Sept–2 October

Juraske R, Mutel C, Stoessel F, Hellweg S (2009) Life cycle human toxicity assessment of pesticides: comparing fruit and vegetable diets in Switzerland and the United States. Chemosphere 77:939–945

Kennedy D, Montgomery D, Quay B (1996) Stochastic environmental life cycle assessment—A probabilistic approach to incorporating variable input data quality. Int J Life Cycle Assess 1(4):199–207

Khazzoom J (1980) Economic implications of mandated efficiency in standards for household appliances. Energy 1(4):21–40

Kim S, Hwang T, Lee KM (1997) Allocation for cascade recycling system. Int J Life Cycle Assess 2(4):217–222

Klöpffer W (1996) Allocation rule for open-loop recycling in life cycle assessment. Int J Life Cycle Assess 1(1):27–31

Klöpffer W (2008) Life cycle sustainability assessment of products (with comments by Helias A. Udo de Haes p.95). Int J Life Cycle Assess 13(2):89–95

Klöpffer W (2012) Verhalten und Abbau von Umweltchemikalien: Physikalisch-chemische Grundlagen, vol 2., völlig überarbeitet. Wiley-VCH, Weinheim

Klöpffer W, Grahl B (2009) Ökobilanz (LCA)—Ein Leitfaden für Ausbildung und Beruf. Wiley-VCH, Weinheim

Knudsen MT, Halberg N (2007) How to include on-farm biodiversity in LCA on food? In: LCA Food Conference, Gothenburg

Koellner T, Scholz R (2008) Assessment of land use impacts on the natural environment. Part 2: Generic characterization factors for local species diversity in Central Europe. Int J Life Cycle Assess 13(1):32–48

Koellner T, Baan L, Beck T, Brandão M, Civit B, Margni M, Canals LMi, Saad R, Souza DMd, Müller-Wenk R (2013) UNEP-SETAC guideline on global land use impact assessment on biodiversity and ecosystem services in LCA. Int J Life Cycle Assess 18:1188–1202. doi:10.1007/s11367-013-0579-z

Koornneef J, Ramirez A, van Harmelen T, van Horssen A, Turkenburg W, Faaij A (2010) The impact of CO_2 capture in the power and heat sector on the emission of SO_2, NO_x, particulate matter, volatile organic compounds and NH_3 in the European Union. Atmos Environ 44(11):1369–1385

Kuswanti C, Xu G, Qiao J, Stuart JA, Koelling K, Lilly B (2003) An engineering approach to plastic recycling based on rheological characterization. J Ind Ecol 6(3–4):125–135

Kuvlesky WP Jr, Brennan LA, Morrison ML, Boydston KK, Ballard BM, Bryant FC (2007) Wind energy development and wildlife conservation—challenges and opportunities. J Wildlife Management 71(8):2487–2498. doi:10.2193/2007-248

Lagerstedt J, Luttropp C, Lindfors L-G (2003) Functional priorities in LCA and design for environment. Int J Life Cycle Assess 8(3):160–166

Landsiedel R, Saling P (2002) Assessment of toxicological risks for life cycle assessment and eco-efficiency analysis. Int J Life Cycle Assess 7(5):261–268

Larsen HF, Olsen SI, Hauschild M, Laurent A (2009) Neptune—new sustainable concepts and processes for optimization and upgrading municipal waster water and sludge treatment/ Deliverable 4.2 Methodology for including specific biological effects and pathogen aspects into LCA. Technical University of Denmark (DTU), Lyngby

Leske T, Buckley c (2003) Towards the development of a salinity impact category for South African environmental life-cycle assessments: Part 1—a new impact category. Water SA 29(3):289–296

Leske T, Buckley C (2004) Towards the development of a salinity impact category for South African life cycle assessments: Part 2– A conceptual multimedia environmental fate and effect model. Water SA 30(2):241–252

Lloyd S, Ries R (2007) Characterizing, propagating, and analyzing uncertainty in life cycle assessment—a survey of quantitative approaches. J Ind Ecol 11(1):161–179

Lo SC, Ma H, Lo SL (2004) Quantifying and reducing uncertainty in life cycle assessment using the Bayesian Monte Carlo method. Science Total Environ 340(1–3):23–33

Luo L, van der Voet E, Huppes G, Udo deHHA (2009) Allocation issues in LCA methodology: a case study of corn stover-based fuel ethanol. Int J Life Cycle Assess 14(6):529–539

Marchand M, Aissani L, Mallard P, Béline F, Réveret J-P (2012) Odour and life cycle assessment (LCA) in waste management—a local assessment proposal. Waste Biomass Valor 4:607-617. doi:10.1007/s12649-012-9173-z

Matthiessen P (2000) Is endocrine disruption a significant ecological issue? Ecotoxicology 9(1–2):21–24

Mattila T, Helin T, Antikainen R, Soimakallio S, Pingoud K, Wessman H (2011) Land use in life cycle assessment. Finnish Environment Institute, Helsinki. doi:http://www.ymparisto.fi/download.asp?contentid=130275&lan=en

Mattila T, Helin T, Antikainen R (2012) Land use indicators in life cycle assessment. A case study on beer production. Int J Life Cycle Assess 17(3):277–286. doi:10.1007/s11367-011-0353-z

May J, Brennan D (2003) Application of data quality assessment methods to an LCA of electricity generation. Int J Life Cycle Assess 8(4):215–225

Meijer A (2007) The significance of indoor air emissions in life cycle assessment of dwellings. International Conference Sustainable Urban Areas, Netherlands

Meijer A, Huijbregts MA, Hertwich E, Reijnders L (2006) Including human health damages due to road traffic in life cycle assessment of dwellings. Int J Life Cycle Assess 11(1):64–71

Michelsen O, Cherubini F, Stromman AH (2012) Impact assessment of biodiversity and carbon pools from land use and land use changes in life cycle assessment, exemplified with forestry operations in Norway. J Ind Ecol 16(2):231–242

Milà i Canals L (2003) Contribution to LCA methodology for agricultural systems. Site-dependency and soil degradation impact assessment., Universitat Autónoma de Barcelona
Milà i Canals L (2007) Land use in LCA—a new subject area and call for papers (editorial). Int J Life Cycle Assess 12(1):1
Milà i Canals L, Bauer C, Depestele J, Dubeuil A, Knuchel R, Gaillard G, Michelsen O, Mueller-Wenk R, Rydgren B (2007a) Key elements in a framework for land use impact assessment within LCA. Int J Life Cycle Assess 12(1):5–15
Milà i Canals L, Romanya J, Cowell S (2007b) Method for assessing impacts on life support functions (LSF) related to the use od fertile land in LCA. J Clean Prod 15(15):1426–1440
Millennium Ecosystem Assessment (2005) Ecosystems and human well-being biodiversity synthesis. World Resources Institute, Washington, DC
Müller NC, Nowack B (2008) Exposure modeling of engineered nanoparticles in the environment. Environ Sci Technol 42:4447–4453
Müller-Wenk R (2002) Attribution to road traffic of the impact of noise on health. Environmental series no 339 Swiss agency for the environment, forests and landscape (SAEFL), Bern
Müller-Wenk R (2004) A method to include in LCA road traffic noise and its health effects. Int J Life Cycle Assess 9(2):76–85
Narščius A, Olenin S, Zaiko A, Minchin D (2012) Biological invasion impact assessment system: from idea to implementation. Ecological Informatics 7(1):46–51. doi:10.1016/j.ecoinf.2011.11.003
National Research Council (2008) Minerals, critical minerals, and the U.S. Economy. National Research Council, Committee on critical mineral impacts of the U.S. economy, Washington, DC
Neugebauer S, Finkbeiner M (2012) The multi-recycling-approach as a new option to deal with end-of-life allocation dilemma. LCA XII—Life cycle thinking … life cycle living, Tacoma
Nishioka Y, Levy JI, Norris GA, Wilson AM, Hofstetter P, Spengler JD (2002) Integrating risk assessment and life cycle assessment—a case study of insulation. Risk Anal 22(5):1003–1017
Noss R (1990) Indicators for monitoring biodiversity— a hierarchical approach. Conserv Biol 4(4):355–364
Notten P, Petrie J (2003) An integrated approach to uncertainty assessment in LCA. In: International workshop on LCI-quality, Karlsruhe, 20–21 October
NRMMC (2006) Australian guideline for water recycling: managing health and environmental risks (phase 1). National water quality management strategy 21. Natural Resource Management Ministerial Council, Environment Protection and Heritage Council, Australian Health Ministers Conference, Canberra
Núñez M, Civit B, Munoz P, Arena AP, Rieradevall J, Anton A (2010) Assessing potential desertification environmental impact in life cycle assessment. Int J Life Cycle Assess 15(1):67–78
Núñez M, Antón A, Muñoz P, Rieradevall J (2013). Inclusion of soil erosion impacts in life cycle assessment on a global scale: application to energy crops in Spain. Int J Life Cycle Assess 18:755–767
NYSERDA (2009) Comparison of reported effects and risks to vertebrate wildlife from six electricity generation types in the New York/New England region. Albany, NY
OECD (2013) Safety of manufactured nanomaterials. Organisation for Economic Co-operation and Development (OECD). http://www.oecd.org/env/ehs/nanosafety/. Accessed 21 May 2013
OED (2013) Oxford dictionairies: “improbable”. http://oxforddictionaries.com/definition/english/improbable?q=improbable+. Accessed 20 Mar 2013
Ou X, Zhang X, Chang S (2010) Scenario analysis on alternative fuel/vehicle for China’s future road transport: Life-cycle energy demand and GHG emissions. J Energy Policy 38(8):3943–3956
PAS 2050 (2008) Specification for the assessment of the life cycle greenhouse gas emissions of goods and services. BSI publications, London
PAS 2050 (2011) Specification for the assessment of the life cycle greenhouse gas emissions of goods and services. BSI publications, London
Pehnt M (2006) Dynamic life cycle assessment (LCA) of renewable energy technologies. Renew Energy 31(1):55–71. doi:http://dx.doi.org/10.1016/j.renene.2005.03.002

Pfister S, Saner D, Koehler A (2011) The environmental relevance of freshwater consumption in global power production. Int J Life Cycle Assess 16(6):580–591

Piemonte V, Gironi F (2011) Land-use change emissions: How green are the bioplastics? Environ Prog Sustain Energy 30(4):685–691. doi:10.1002/ep.10518

Piemonte V, Gironi F (2012) Bioplastics and GHGs saving: the land use change (LUC) emissions issue. Energ Source Part A 34(11):1995–2003. doi:10.1080/15567036.2010.497797

Plevin RJ, O'Hare M, Jones AD, Torn MS, Gibbs HK (2010) Greenhouse gas emissions from biofuels' indirect land use change are uncertain but may be much greater than previously estimated. Environ Sci Technol 44(21):8015–8021. doi:10.1021/es101946t

Podmore C (2009) Irrigation salinity causes and impacts. Department of Primary Industries. DPI http://www.dpi.nsw.gov.au/__data/assets/pdf_file/0018/310365/Irrigation-salinity-causes-and-impacts.pdf. Accessed 29 May 2013

Pope C, Burnett RT, Krewski D, Jerrett M, Shi Y, Calle EE, Thun MJ (2009) Cardiovascular mortality and exposure to airborne fine particulate matter and cigarette smoke—shape of the exposure-response relationship. Circulation 120(11):941–948

Potting J, Preiss P, Seppala J, Struijs J, Wiertz J, Blazek M, Heijungs R, Itsubo N, Masanet E, Nebel B, Nishioka Y, Payet J, Becaert V, Basset-Mens C, Jolliet O (2007) Current practice in LCIA of transboundary impact categories. Report of task force 4 on transboundary impacts. UNEP/SETAC Life Cycle Initiative. Life cycle impact assessment programm

Pro Igel (2012) Wildtier Igel. Eine kleine Igelkunde. www.pro-igel.de/merkblaetter/publpdfs/wildtier.pdf. Accessed 8 Feb 2013

Rabl A, Benoist A, Dron D, Peuportier B, Spadaro JV, Zoughaib A (2007) How to account for CO_2 emissions from biomass in an LCA. Int J Life Cycle Assess 12(5):281

Reap J, Bras B, Newcomb PJ, Carmichael C (2003) Improving life cycle assessment by including spatial, dynamic and place-based modeling. Paper no DETC2003/DFM-48140. In: ASME 2003 International design engineering technical conferences and computers and information in engineering conference, Chicago, Illinois, 2–6 September

Reap J, Roman F, Duncan S, Bras B (2008a) A survey of unresolved problems in life cycle assessment, Part 1: goal and scope and inventory analysis. Int J Life Cycle Assess 13(4):290–300

Reap J, Roman F, Duncan S, Bras B (2008b) A survey of unresolved problems in life cycle assessment, Part 2: impact assessment and interpretation. Int J Life Cycle Assess 13(5):374–388

Reimann K, Finkbeiner M, Horvath A, Matsuno Y (2010) Evaluation of environmental life cycle approaches for policy and decision making support in micro and macro level applications. Institute for Environment and Sustainability, Joint Research Centre, European Commission

Reinhard J, Zah R (2009) Global environmental consequences of increased biodiesel consumption in Switzerland—consequential life cycle assessment. J Clean Prod 17:46–56

Risku-Norja H, Mäenpää I (2007) MFA model to assess economic and environmental consequences of food production and consumption. Ecolog Econ 60(4):700–711. doi:10.1016/j.ecolecon.2006.05.001

Rosenbaum RK, Bachmann TM, Gold LS, Huijbregts MAJ, Jolliet O, Juraske R, Koehler A, Larsen HF, MacLeod M, Margni M, McKone TE, Payet J, Schuhmacher M, van de Meent D, Hauschild MZ (2008) USEtox-the UNEP-SETAC toxicity model: recommended characterisation factors for human toxicity and freshwater ecotoxicity in life cycle impact assessment. Int J Life Cycle Assess 13(7):532–546

Rosenbaum R, Huijbregts MJ, Henderson A, Margni M, McKone T, Meent D, Hauschild M, Shaked S, Li D, Gold L, Jolliet O (2011) USEtox human exposure and toxicity factors for comparative assessment of toxic emissions in life cycle analysis: sensitivity to key chemical properties. Int J Life Cycle Assess 16(8):710–727. doi:10.1007/s11367-011-0316-4

Rosenbaum RK, Owsianiak M, Hauschild MZ, Vijver MG, Heijungs R, Peijnenburg WJGM, Golsteijn L, Zelm Rv, Musters G, Hendriks AJ, Huijbregts MAJ, Fantke P, Juraske R, Anton A, Itoiz ES, Kounina A, Humbert S, Jolliet O, Margni M, Sala S, Marinov D, Pennington D (2012) Recommended assessment framework, method and characterisation factors for ecotoxicity and human toxicity: phase 1 (report and model+factors). LC-IMPACT, vol Version 3. Worpackage 2, DTU

Rugani B, Panasiuk D, Benetto E (2012) An input–output based framework to evaluate human labour in life cycle assessment. Int J Life Cycle Assess 17(6):795–812. doi:10.1007/s11367-012-0403-1

Saad R, Margni M, Koellner T, Wittstock B, Deschenes L (2011) Assessment of land use impacts on soil ecological functions: development of spatially differentiated characterization factors within a Canadian context. Int J Life Cycle Assess 16(3):198–211. doi:10.1007/s11367-011-0258-x

Sanchez ST, Woods J, Akhurst M, Brander M, O'Hare M, Dawson TP, Edwards R, Liska AJ, Malpas R (2012) Accounting for indirect land-use change in the life cycle assessment of biofuel supply chains. J R Soc Interface 9(71):1105–1119. doi:10.1098/rsif.2011.0769

Schenck RC (2001) Land use and biodiversity indicators for life cycle impact assessment. Int J Life Cycle Assess 6(2):114–117

Schettkat R (2009) Analyzing rebound effects. Schumpeter discussion papers, No 2009-002. Schumpeter school of business and economics, University of Wuppertal, Wuppertal

Schleisner L (2000) Life cycle assessment of a wind farm and related externalities. Renew Energy 20(3):279–288. doi:http://dx.doi.org/10.1016/S0960-1481(99)00123-8

Schmidt JH, Weidema BP (2008) Shift in the marginal supply of vegetable oil. Int J Life Cycle Assess 13(3):235–239

Schneider L, Berger M, Finkbeiner M (2011a) The anthropogenic stock extended abiotic depletion potential (AADP) as a new parameterisation to model the depletion of abiotic resources. Int J Life Cycle Assess 16(9):929–936

Schneider L, Berger M, Finkbeiner M (2011b) Economic material availability as a new area of protection for life cycle sustainability assessment. In: SETAC Europe 21st Annual Meeting, 15–19 May, Milano

Schneider L, Berger M, Finkeiner M (2012) Measuring material scarcity—limited availability despite sufficient reserves. Paper presented at the European Commission Joint Research Centre Workshop: Security of supply and scarcity of raw materials: a methodological framework for supply chain sustainability assessment, November 13–14, Ranco

Schneider L, Berger M, Schüler-Hainsch E, Knöfel S, Ruhland K, Mosig J, Bach V, Finkbeiner M (2013) The economic resource scarcity potential (ESP) for evaluating resource use based on life cycle assessment. Int J Life Cycle Assess, published online, DOI 10.1007/s11367-013-0666-1

Seager TP, Linkov I (2008) Coupling multicriteria decision analysis and life cycle assessment for nanomaterials. J Ind Ecol 12(3):282–285. doi:10.1111/j.1530-9290.2008.00048.x

Shiklomanov IA (2003) World water resources at the beginning of the 21st century. International hydrology series. Cambridge University Press, Cambridge

Sioutas C, Delfino RJ, Singh M (2005) Exposure assessment for atmospheric ultrafine particles (UFPs) and implications in epidemiologic research. Environ Health Perspectives 113(8):947–955

Smallwood KS, Thelander CG, Morrison ML, Rugge LM (2007) Burrowing owl mortality in the Altamont Pass Wind Resource Area. J Wildlife Management 71(5):1513–1524. doi:10.2193/2006-307

Som C, Berges M, Chaudhry Q, Dusinska M, Fernandes TF, Olsen SI, Nowack B (2010) The importance of life cycle concepts for the development of safe nanoproducts. Toxicology 269(2–3):160–169. doi:10.1016/j.tox.2009.12.012

Sonnemann G, Schuhmacher M, Castells F (2003) Uncertainty assessment by a Monte Carlo simulation in a life cycle inventory of electricity produced by a waste incinerator. J Clean Prod 11(3):279–292

Souza D, Flynn D, de Clerck F, Rosenbaum R, de Melo LH, Koellner T (2013) Land use impacts on biodiversity in LCA: proposal of characterization factors based on functional diversity. Int J Life Cycle Assess 18:1231–1242. doi: 10.1007/s11367-013-0578-0

Spadaro JV (2004) RiskPoll: a model for estimating public health and environmental impacts of air pollution. Center for energy efficiency and systems, http://www.ces.mines-paristech.fr/english/themes/impact/software/RiskPoll%20Overview.pdf. Accessed 28 May 2013

Spielmann M, Haan P, Scholz RW (2008) Environmental rebound effects of high-speed transport technologies: a case study of climate change rebound effects of a future underground maglev train system. J Clean Prod 16(13):1388–1398
Steen B (1999) A systematic approach to environmental priority strategies in product development (EPS). Version 2000—Models and data of the default method. CPM report 1999:5. Centre for Environmental Assessment of Products and Material Systems, Chalmers University of Technology, Environmental Systems Analysis
Strømman AH, Hertwich E (2004) Hybrid life cycle assessment of large scale hydrogen production facilities. Working papers no3/2004. Norwegian University of Science and Technology (NTNU), Industrial Ecology Programme (IndEcol), Trondheim
Suh S, Lenzen M, Treloar GJ, Hondo H, Horvath A, Huppes G, Jolliet O, Klann U, Krewitt W, Moriguchi Y, Munksgaard J, Norris GA (2004) Critical review system boundary selection in life-cycle inventories using hybrid approaches. Environ Sci Technol 38(3):657–664
Suh S, Nakamura S (2007) Editorials five years in the area of input-output and hybrid LCA. Int J Life Cycle Assess 12(6):351–352
Swarr T, Hunkeler D, Klöpffer W, Pesonen HL, Ciroth A, Brent AC, Pagan R (2011) Environmental life cycle costing—a code of practice. Int J Life Cycle Assess 16:389–391
Thomassen MA, Dalgaard R, Heijungs R, de Boer I (2008) Attributional and consequential LCA of milk production. Int J Life Cycle Assess 13(4):339–349
Thomassen M, Dolman M, van Calker KJ, de Boer J (2009) Relating life cycle assessment indicators to gross value added for Dutch dairy farms. Ecolog Econ 68(8–9):2278–2284
Tukker A (2002) Risk analysis, life cycle assessment—The common challenge of dealing with the precautionary frame (based on the toxicity controversy in Sweden and The Netherlands). Risk Anal 22(5):821–832
Udo de Haes H A (1993) Applications of life cycle assessment: expectations, drawbacks and perspectives. J Clean Prod 1(3–4):131–137
UNEP/SETAC (2009) Guidelines for social life cycle assessment of products. Druk in deweer, Belgium
UNEP/SETAC (2011a) Global guidance principles for life cycle assessment databases. A basis for greener processes and products. UNEP/SETAC Life Cycle Initiative, Paris
UNEP/SETAC (2011b) Towards a life cycle sustainability—making informed choices on products. Druk in de weer, Belgium
UNEP—United Nations Environment Programme (2010a) Assessing the environmental impacts of consumption and production; priority products and materials. A report of the working group on the environmental impacts of products and materials to the international panel for sustainable resource management
UNEP—United Nations Environment Programme (2010b) Connecting the dots—Biodiversity, adaptation, food security and livelihoods
UN—United Nations (1992) Convention on biological diversity. Secretariat of the Convention on biological diversity, http://www.cbd.int/doc/legal/cbd-en.pdf. Accessed 28 May
UN—United Nations (2011) Convention to combat desertification: a visual synthesis. UNCCD, http://www.unccd.int/Lists/SiteDocumentLibrary/Publications/Desertification-EN.pdf. Accessed 29 May 2013
US EPA (2012) Endocrine disrupter screening programme (EDSP). United States Environmental Protection Agency
US Green Building Council (2008) Getting LCA into LEED: a backgrounder on the first LCA pilot credit for LEED
van der Voet E, de Bruyn S, Tukker A (2009) Environmental impact of the use of natural resources and products. Leiden
van der Voet E, Lifset RJ, Luo L (2010) Life-cycle assessment of biofuels, convergence and divergence. Biofuels 1(3):435–449
van Kempen EEMM, Kruize H, Boshuizen HC, Ameling CB, Staatsen BAM, de Hollander AEM (2002) The association between noise exposure and blood pressure and ischemic heart disease: a meta-analysis. Environ Health Perspectives 110(3):307–317

van Zelm R, Huijbregts MAJ, den Hollander HA, van Jaarsveld HA, Sauter FJ, Struijs J, van Wijnen HJ, van deMD (2008) European characterization factors for human health damage of PM10 and ozone in life cycle impact assessment. Atmos Environ 42(3):441–453

Vandenberg LN, Colborn T, Hayes TB, Heindel JJ, David R. Jacobs, Lee D-H, Shioda T, Soto AM, Saal FSv, Welshons WV, Zoeller RT, Myers JP (2012) Hormones and endocrine-disrupting chemicals: low-dose effects and nonmonotonic dose responses. Endocr Rev 33(3):378–455

Varun, Bhat IK, Prakash R (2009) LCA of renewable energy for electricity generation systems—a review. Renew Sust Energy Rev 13(5):1067–1073. doi:http://dx.doi.org/10.1016/j.rser.2008.08.004

VDI (2013) http://www.vdi.de/technik/fachthemen/energie-und-umwelt/fachbereiche/ressourcen-management/themen/richtlinienwerk-zur-ressourceneffizienz-zre/. Accessed 7 January 2013

Vogtländer JG, Lindeijer E, Witte JPM, Hendriks C (2004) Characterizing the change of land use based on flora: application for EIA and LCA. J Clean Prod 12(1):47–57

Watson RT, Rosswall T, Steiner A, Töpfer K, Arico S, Bridgewater P (2005) Ecosystems and human well-being: biodiversity sysnthesis. In: Watson RT, Rosswall T, Steiner A, Töpfer K, Arico S, Bridgewater P (eds) Ecosystems. World Resources Institute, Washington. doi:10.1196/annals.1439.003

Watts BD (2010) Wind and waterbirds—Establishing sustainable mortality limits within the Atlantic flyway. Williamsburg

Weidema BP, Wesnæs M (1997) Data quality management for life cycle inventories—an example of using data quality indicators. J Clean Prod 4(3–1):167–174

Weidema BP, Wesnaes M, Hermansen J, Kristensen T, Halberg N (2008) Environmental improvement potentials of meat and dairy products (EUR 23491 EN). European Commission, Joint research centre (JRC), Institute for prospective technological studies (IPTS), Sevilla

Werner F, Richter K (2000) Economic allocation in LCA—a case study about aluminumg window frames. Int J Life Cycle Assess 5(2):79–83

Wischmeier W, Smith D (1978) Predicting rainfall erosion losses—a guide to conservation planning. EPA, http://topsoil.nserl.purdue.edu/usle/AH_537.pdf. Accessed 29 May 2013

Wong CS, Green DR, Cretney WJ (1974) Quantitative tar and plastic waste distributions in the Pacific ocean. Nature 247(4):30–32

Wötzel K (2007) Ökobilanzierung der Altfahrzeugverwertung am Fallbeispiel eines Mittelklasse-fahrzeuges und Entwicklung einer Allokationsmethodik. Technische Universität Braunschweig, Braunschweig

WRI/WBCSD (2004) The greenhouse gas protocol—a corporate accounting and reporting standard. USA

Yee KF, Tan KT, Abdullah AZ, Lee KT (2009) Life cycle assessment of palm biodiesel: revealing facts and benefits for sustainability. Appl Energy 86(Supplement 1):189–196

Zhai P, Williams ED (2010) Dynamic hybrid life cycle assessment of energy and carbon of multi-crystalline silicon photovoltaic systems. Environ Sci Technol 44(20):7950–7955

Index

W. Klöpffer (ed.), *Background and Future Prospects in Life Cycle Assessment*,
LCA Compendium – The Complete World of Life Cycle Assessment,
DOI 10.1007/978-94-017-8697-3, © Springer Science+Business Media Dordrecht 2014

MIX
Papier aus verantwortungsvollen Quellen
Paper from responsible sources
FSC® C105338
FSC
www.fsc.org

If you have any concerns about our products,
you can contact us on
ProductSafety@springernature.com

In case Publisher is established outside the EU,
the EU authorized representative is:
Springer Nature Customer Service Center GmbH
Europaplatz 3, 69115 Heidelberg, Germany

Printed by Libri Plureos GmbH
in Hamburg, Germany